WORLD SOILS

E. M. BRIDGES

Lecturer in Geography at
University College of
Swansea

WORLD SOILS

CAMBRIDGE · AT THE UNIVERSITY PRESS · 1970

PUBLISHED BY
THE SYNDICS OF THE CAMBRIDGE UNIVERSITY PRESS
Bentley House, 200 Euston Road, London N.W.1
American Branch: 32 East 57th Street, New York, N.Y. 10022

Library of Congress Catalogue Card Number: 76-96082
International Standard Book Number: 0 521 07616 1 school edition
0 521 08047 9 cloth edition

Printed in Great Britain
by Jarrold and Sons Ltd, Norwich

CONTENTS

ACKNOWLEDGEMENTS

The author wishes to acknowledge the help given by the following: Miss G. Thomas for drawing the diagrams, Mr G. Lewis for compiling the world soil map, Miss N. Thomas and Miss E. Williams for secretarial assistance.

Thanks are due to D. E. Cotton for permission to reproduce Plate 1, to B. Clayden for Plate 11, to A. Young for Plate 19 and to D. Gunary for Plate 25. The other colour plates are by the author.

For permission to reproduce diagrams and photographs, thanks are due to the following: S. R. Eyre for Figs. 1.1 and 2.7 from *Vegetation and Soils*, Edward Arnold, 1969; H. O. Buckman and N. C. Brady for Figs. 2.1, 2.4b and 2.9 from *The Nature and Properties of Soils*, Macmillan, 1960; U.S.D.A. for Figs. 2.2 and 2.8 from Soil Survey Staff, *Soil Survey Manual*, 1951 and for Fig. 3.9 (a photo by W. M. Johnson) from *Soil Classification, A Comprehensive System 7th Approximation*, 1960; G. P. C. Chambers for Fig. 2.4a from 'Natural Colloidal Silicates' in *Science News* 40, Penguin Books, 1956; P. Duchafour for Fig. 2.5 from *Précis de Pédologie*, Masson et Cie, 1965; E. J. Russell for Fig. 3.3 from *The World of the Soil*, Collins Fontana Library, 1961; G. R. Clarke for Fig. 5.3 from *The Study of Soil in the Field*, Clarendon Press, 1961; Van Riper for Fig. 6.1 from *Man's Physical World*, McGraw Hill; Australian C.S.I.R.O. for Fig. 5.4 from R. A. Perry *et al.*, *General Report on Lands of the Alice Springs Area 1956–7*, 1962 and for Fig. 9.1 from *Periodic Phenomena in Landscapes as a Basis for Soil Studies*, 1959; A. K. Lobeck for Fig. 8.6 from *Geomorphology*, McGraw Hill; B. T. Bunting for Fig. 10.8 from *The Geography of the Soil*, Hutchinson University Library, 1965. Fig. 5.1 is partly based upon a map which appeared in *Efficient Use of Fertilizers*, F.A.O., 1958 and includes modifications derived from Ganssen and Hädrich *Atlas zur Bodenkunde*, Bibliographisches Institute Mannheim, Kartographisches Institute Meyer, 1965.

Soil profile descriptions quoted as examples in the text are taken from the following sources:

Afanasyeva, E. A., *et al.*, 1964, *Short Guide to Soil Excursion Moscow-Kherson* VIII Int. Cong. Soil Sci., Ministry of Agriculture, Moscow (descriptions by E. A. Afanasyeva, V. M. Fridland, G. S. Grin and V. D. Kissel); Tours Guide, Australian Soil Science Conference, 1966, Brisbane; Aubert, G. and Boulaine, J., 1967, *La Pédologie. Que Sais-je?*. Presses Universitaires de France, Paris (descriptions by G. Aubert and N. Federoff); Bleakley, D. and Khan, E. J. A., 1963, 'Observations on the white sand areas of the Berbice Formation', British Guiana, *J. Soil Sci.*, 14, 44–51; Bridges, E. M., 1966, *The Soils and Land Use of the District north of Derby*, Mem. Soil Survey of Great Britain, Harpenden; D'Hoore, J. L., 1964, *Soil Map of Africa*, 1:5,000,000, C.T.C.A., Lagos (descriptions by J. V. Botelho da Costa, J. H. Durand, R. Frankart, J. Hervieu, N. Leneuf and G. Riou, R. Maignien and C. R. Van der Merwe); Mackney, D. and Burnham, C. P., 1964, *The Soils of the West Midlands*, Bull. no. 2, Soil Survey of Great Britain, Harpenden; Mückenhausen, E., 1956, 'Typologische Bodenentwicklung und Bodenfruchtbarkeit', pp. 37–103, in *Arbeitsgemeinschaft fur Forschung des Lands Nordrhein-Westfalen*, Westdeutscher Verlag, Köln; Mulcahy, M., 1960, 'Laterites and Lateritic soils in Southwestern Australia', *J. Soil Sci.*, 11, pp. 206–25; Northcote, K. H., *et al.*, 1954, *Soils and Land Use in the Barossa District, South Australia*, Soils and Land Use Series No. 13, C.S.I.R.O., Melbourne; Oakes, H., 1954, *The Soils of Turkey*, Ministry of Agriculture, Ankara; Soil Survey Staff, 1951, 'Soil Survey Manual', *Agricultural Handbook No. 18*, U.S.D.A., Washington; Svatkov, N. M., 1958, 'Soils of Wrangel Island', *Soviet Soil Science*, pp. 80–7; Tedrow, J. C. F. and Hill, D. E., 1955, 'Arctic Brown Soil', *Soil Science* 80, pp. 265–75; Thorp, J., 1957, *Report on a Field Study of Soils of Australia*, Earlham College, Richmond, Indiana (Mimeo); Zalibekov, Z. G., 1965, 'Separation of cinnamon-brown soils on the Aktashsk Sub-montane Dagestan Plain', *Soviet Soil Science*, pp. 1158–65.

1 INTRODUCTION

Soil has a peculiar fascination, which impinges upon all of us at some time or other. Farmers or horticulturists till it, engineers move it about in Juggernaut-like machines, small boys dig in it, and mothers abhor it as being dirty. Unfortunately, for many people soil is synonymous with dirt. They should know better, for the soil has a vital and important role to play in the life of the world and mankind. As Sir John Russell has written, 'a clod of earth seems at first sight to be the embodiment of the stillness of death', however he goes on to show that it is in fact a highly organised physical, chemical and biological complex on which all of us are dependent. As the supporter of vegetable life, the soil plays the most fundamental of roles in providing sustenance for all animals and man.

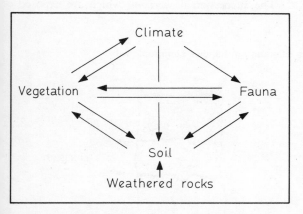

1.1 The arrows indicate the various interactions of the environment which affect the soil

The position of soil in the diagram of the biotic complex shows that climate influences plants, animals and soil directly (Fig. 1.1). Plants influence the soil, the animals and the climate near the ground. Animals play a considerable role in soil development, the type of soil often influences the animals which are present in it, while the animals also influence the vegetation which is growing in the soil. Finally climate, through weathering, influences the rocks, which in time become part of the soil as it is first weathered and later acted upon by soil-forming processes.

The study of soils is the occupation of the *pedologist*, and his science of *pedology* emphasises the study of the soil as a natural phenomenon on the surface of the earth. The pedologist is interested in the appearance of the soil, its mode of formation, its physical, chemical and biological composition, and its classification and distribution.

Pedology makes use of a large number of branches of scientific knowledge, and as an integrative science resembles the role of geography. As will already be apparent from the previous paragraph, aspects of physics, chemistry and biology have an important contribution to the study of the soil, as have also studies in agriculture, forestry, history, geography, mineralogy, archaeology and geology. From all of these are obtained information which can be synthesised to make a scientific discipline and natural philosophy separate from, and yet closely related to, many other branches of natural science.

Pedology can be studied as a pure science in which the identification of the processes producing the soil profile, as well as the mapping and classification of soils, form an important part. However, the results obtained in the pure science can be applied to practical problems in agriculture, horticulture, forestry, engineering, and in planning the future use of the land.

In cases of the proposed development of virgin lands or of lands previously used for extensive grazing, the pedologist can offer recommendations for the cultivation practices and the parcelling out of farm units within the developing area. This type of work is particularly useful in irrigation schemes.

The high cost of installation makes a knowledge of the soils essential before the civil engineering work is even planned. The classification of soils leads naturally to land capability, hence the pedologist's interest in the natural fertility of soils, their best use or even possible increase of fertility.

Definition of soil

Present-day soil science has emerged from two different schools of thought, one chemical, the other geological. The German scientist Liebig was probably the most renowned exponent of the chemical view of the soil, but even before Liebig, a Swedish scientist, Berzelius, described soil as 'the chemical laboratory of nature in whose bosom various chemical decomposition and synthesis reactions take place in a hidden manner'. Early pedologists with a geological background considered the soil to be comminuted rock with a certain amount of organic matter derived from the decomposition products of plants. As late as 1917, a German scientist, Ramann, described soil as 'rocks that have been reduced to small fragments and have been more or less changed chemically, together with the remains of plants or animals that live in it or on it'.

Current definitions of soil result mainly from the work of two men, in Russia, Dokuchaiev, and in America, Hilgard. Independently both noted that soils were related in a general way to climate, and that soils could be described in broad geographical zones, which at the scale of world maps could be correlated not only with climate but also with the associated belts of vegetation. Although this is only partly true, it did serve to direct attention to the environmental relationships of the soil cover of our planet. This environmental approach still holds true today. The definition of soil propounded by Joffe has the advantage that it combines the physical, chemical and biological constituents, and throws the right amount of weight on the importance of morphology in the description of a soil.

'The soil is a natural body of animal, mineral and organic constituents differentiated into horizons of variable depth which differ from the material below in morphology, physical make-up, chemical properties and composition, and biological characteristics.' A more simple definition states that soil is 'the stuff in which plants grow', however this is not necessarily a full definition. Everyone knows cress will grow on a damp piece of flannel without the aid of soil, and by the science of hydroponics plants are grown in water with the necessary mineral nutrients.

The succession of *horizons* which are exposed

1.2 An area of country with its soil profile; expanded section on right shows soil profile with A, E, B and C horizons

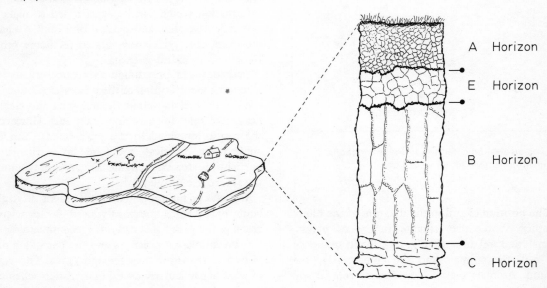

A Horizon

E Horizon

B Horizon

C Horizon

when a vertical cut is made through the soil to the parent material comprises the *soil profile* (Fig. 1.2). Where present, superficial organic horizons are designated by the letters L, F and H; surface eluvial horizons, A and E; sub-surface illuvial horizons by B, whilst the parent material is indicated by the letter C (Fig. 5.6). The appearance of the horizons and their order can supply much information on the developmental processes which are operating in the soil. These processes are discussed in Chapter 3 following a discussion of the composition of soils in Chapter 2. The soil itself is normally developed in material which has already been weathered from the solid rocks. This *weathered mantle* or *regolith* can be as deep as 50 m. in the humid tropics. In Britain the depth is variable, but on average is somewhere around 1·5 m. although in places it is non-existent.

Before attempting a regional account of the soils of the world, it is necessary to know something of the ways whereby soils are classified and mapped. This is dealt with in Chapters 4 and 5 which conclude the first section of this book.

The zonal soils of the world are described in Chapters 6 to 9 and a further discussion of intra-zonal and azonal soils is included in Chapter 10. Examples of soil profiles from published and unpublished sources are given as illustrations of the different soil types. The original author's description has been retained where possible, but some simplification and rearrangement has often been necessary. This accounts for the variability of the different descriptions. A discussion of the part soil science has to play in the determination of the best use of our soils for the human race as a whole concludes the book.

2 COMPOSITION OF SOILS

There are four main constituents of soil: mineral matter, organic matter, air and water (Fig. 2.1). The mineral matter includes all those minerals weathered from the parent material as well as those formed in the soil by recombination from substances in the soil solution. The organic matter is derived mostly from decaying vegetable matter which is broken down and decomposed by the action of the many different forms of animal life which live in the soil. Normally both air and water occupy the spaces between the structures of the soil, but if a soil is saturated with water most of the air is driven out. In a soil which is freely drained some water is still present in the form of thin films around the mineral particles, leaving the spaces of the fissures and pores open for the penetration of the atmosphere.

of less than 0·002 mm. diameter to sand-size particles of up to 2 mm. diameter. This part of the soil is known as the *fine earth* and it is upon this that the *texture* of the soil is determined. Larger particles or stones occur also, but except for their bulk are considered to be inert, contributing only by their physical presence. This can be useful in a fine-textured soil in that the stones break the continuity of the clay material. Taking the three different fractions of sand, silt and clay which occur in any soil, it is possible to relate them to the triangular diagram (Fig. 2.2). The texture of a soil can easily be determined in the field by first moistening and then estimating the proportions of sand, silt and clay as it is worked between finger and thumb. (Sand 2 mm.–0·05 mm., silt 0·05–0·002 mm., clay <0·002 mm.) Descriptions of the different classes of soil texture are given in Table 2.1.

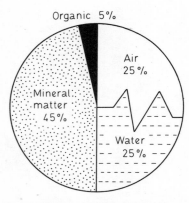

2.1 Volume composition of a typical topsoil: amounts are approximate as the percentage of certain constituents, e.g. air and water, is constantly varying

Mineral matter

The mineral portion of soil is derived from the parent material by weathering and consists of a range of particle sizes from very small clay particles

Table 2.1 *Soil texture class descriptions*

Sand. Soil consisting mostly of coarse and fine sand, and containing so little clay that it is loose when dry and not sticky at all when wet. When rubbed it leaves no film on the fingers.

Loamy sand. Consisting mostly of sand, but with sufficient clay to give slight plasticity and cohesion when very moist. Leaves a slight film of fine materials on the fingers when rubbed.

Sandy loam. Soil in which the sand fraction is still quite obvious, which moulds readily when sufficiently moist, but in most cases does not stick appreciably to the fingers. Threads do not form easily.

Loam. Soil in which the fractions are so blended that it moulds readily when sufficiently moist, and sticks to the fingers to some extent. It can with difficulty be moulded into threads but will not bend into a small ring.

Silt loam. Soil that is moderately plastic without being very sticky and in which the smooth, soapy feel of the silt is the main feature.

Sandy clay loam. Soils containing sufficient clay to be distinctly sticky when moist, but in which the sand fraction is still an obvious feature.

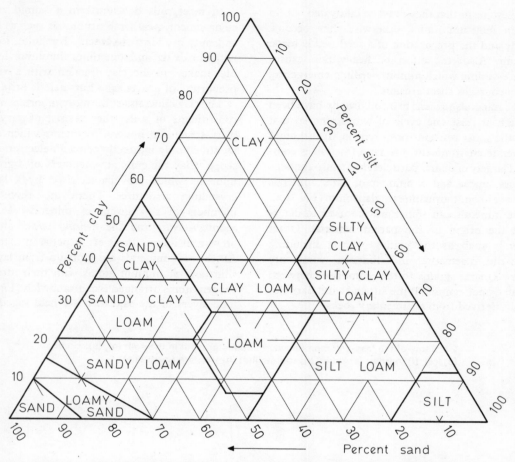

2.2 Soil texture. The three sides represent base lines for sand, silt and clay with the apices opposite representing 100 per cent of each constituent. Percentages can be read off to give the textural name for any soil sample

Clay loam. The soil is distinctly sticky when sufficiently moist, and the presence of sand fractions can only be detected with care.

Silty clay loam. This contains quite subordinate amounts of sand, but sufficient silt to confer something of a smooth, soapy feel. It is less sticky than silty clay or clay loam.

Silt. Soil in which the smooth, soapy feel of silt is dominant.

Sandy clay. The soil is plastic and sticky when moistened sufficiently, but the sand fraction is still an obvious feature. Clay and sand are dominant and the intermediate grades of silt and very fine sand are less apparent.

Clay. The soil is plastic and sticky when moistened sufficiently and gives a polished surface on rubbing. When moist the soil can be rolled into threads, and it is capable of being moulded into any shape and takes clear fingerprints.

Silty clay. Soil which is composed almost entirely of very fine material, but in which the smooth, soapy feel of the silt fraction modifies to some extent the stickiness of the clay.

These different soil textures have properties which affect the management and economic use of the soil. Coarse-textured, sandy soils are usually freely drained, and in a dry summer may suffer from drought, but cultivation is relatively easy. Frequently, clay soils are poorly drained, and the expense of installing a drainage system can be large. Cultivations are always likely to be difficult, although in a dry year these soils may produce better crops than a sandy soil. Silty soils are also troublesome in that they must be cultivated within certain moisture limits, otherwise they become cloddy and the preparation of a seed-bed is made difficult. Also, the effect of heavy rain causes surface-sealing which inhibits seedling emergence, and encourages sheet erosion.

The minerals present in a soil usually have been through at least one cycle of weathering so that only the most resistant ones remain. In a humid temperate environment, the sand fraction is composed largely of quartz particles, but it also contains felspars, micas and a number of rarer minerals such as zircon, tourmaline or glauconite (Fig. 2.3). These minerals can sometimes be used to determine the origin of the parent material. Other minerals such as the oxides of iron including magnetite, haematite and limonite commonly occur. Quartz grains often comprise between 90 and 95 per cent of all the silt and sand particles in soils derived from sedimentary rocks.

In most soils developed in a humid tropical environment, weatherable minerals such as felspar and mica are virtually absent. Kaolinite, together with iron oxides and sometimes aluminium hydroxide, makes up the clay fraction with a variable proportion of quartz sand but there is little silt.

The clay minerals are the most important mineral constituents of soils; they consist of very small platy-structured mineral fragments which can be identified only indirectly or by an electron microscope. Clay minerals are members of a group of minerals which are characterised by a layered, crystalline structure. There are three main members of this group of minerals, *kaolinite*, *montmorillonite* and the *hydrous micas*, although other transitional forms occur between each type. All the clay minerals are built up from layers of silica and aluminium atoms with their attendant oxygen atoms arranged like a sandwich (Fig. 2.4). The layers of the sandwich are held together by

2.3 Some of the less-common (heavy) minerals present in the soil: (a), (b), zircon; (d), (e), (f), garnet; (c) rutile; (g) tourmaline; (h) iron ore (magnetite)

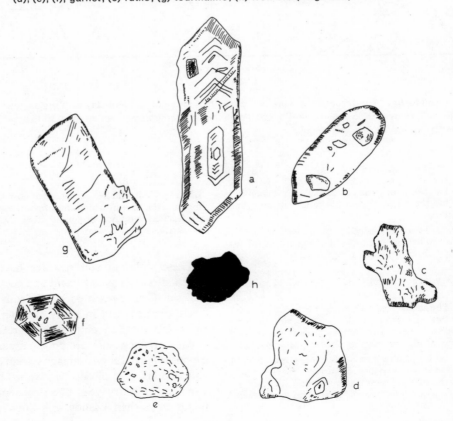

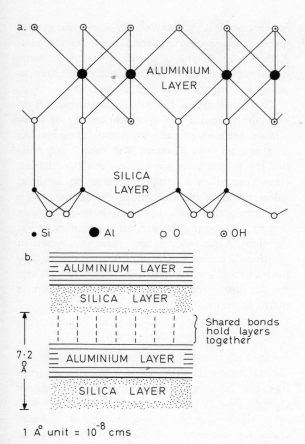

a.

ALUMINIUM
LAYER

SILICA
LAYER

● Si   ● Al   ○ O   ⊙ OH

b.

═ ALUMINIUM LAYER ═

SILICA   LAYER

Shared bonds
hold layers
together

7·2 Å

ALUMINIUM   LAYER

SILICA   LAYER

$1 \text{ Å unit} = 10^{-8} \text{ cms}$

2.4 (a) Arrangement of the atoms in the clay mineral
kaolinite; (b) Diagrammatic representation of the
layered structure typical of the clay minerals

chemical bonds shared between the different
layers.

Some minerals are formed in the soil itself.
Weathering releases elements from the minerals
and they pass into the soil solution from which
recrystallisation takes place forming new minerals.
Hence kaolinite can be formed in this way from
soil solutions rich in aluminium and silicon.
Where base-rich conditions prevail the mineral
montmorillonite may be formed. Clay minerals
can also be formed in the soil by alteration of the
primary minerals. Other secondary minerals can
accumulate in the soil. Where gleying is prevalent
concretions of iron and manganese oxides occur
scattered throughout the sub-soil horizons of some
surface-water gley soils. These grow by the
addition of concentric layers of iron and manganese
compounds. Concretions of calcareous material are

unusual in Britain and other humid countries
because of the leaching which takes place, but they
are common in the soils of drier climates such as in
the chernozems of the steppes of Russia.

Because the clay mineral particles are so small in
size, the minute electrical forces of the molecules
at the surface of the clay become dominant and
confer upon the clay particles a *colloidal* status. A
colloidal state occurs when particles of less than
one micron (0·001 mm.) in size are dispersed
evenly throughout another material. (Two common
examples of colloids are milk, in which tiny solid
particles are dispersed throughout a liquid, and
cloud, where water droplets are dispersed throughout a gaseous mixture.) Properties which are
conferred upon a soil by the colloidal state are
plasticity, cohesion, shrinkage, swelling, flocculation and dispersion (Fig. 2.5).

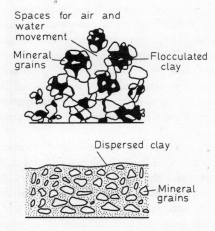

Spaces for air and
water
movement

Mineral
grains

Flocculated
clay

Dispersed clay

Mineral
grains

2.5 The effect of flocculation and dispersion on soil.
The flocculated soil is well-structured, and has spaces
through which air and water can move. These spaces
are lost or very much reduced when a soil is
dispersed

Organic matter

Soil organic matter can take several forms; it
may be intimately mixed with the mineral matter
of the surface horizon, it may be present in an
illuvial horizon, or it may lie upon the surface.
The four main types of surface organic matter are
known as *mull*, *moder*, *mor* and *peat* (Fig. 2.6).
Gradational forms used by soil scientists can be
recognised between these types, though they
remain closely related to the basic varieties.

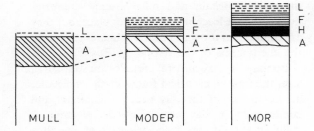

2.6 Diagrammatic representation of the thin surface forms of organic matter accumulation in soils

Mor develops best beneath a heathland or coniferous forest plant community. As the litter, L, which falls from the plants is low in base content, strong acidity rapidly develops which inhibits the activity of the soil fauna. Breakdown is retarded so that the litter accumulates to form a transition or fermentation F horizon with increasingly matted material towards the base of the horizon. A thin horizon of completely humidified and amorphous material forms the humus H horizon. As the activity of the soil fauna is restricted in this type of organic matter, much of the breakdown is achieved by the action of fungi. Earthworms are usually absent so that incorporation of the humus into the mineral soil is extremely slow.

Moder is a form of organic matter intermediate between mor and mull, having a richer soil fauna than mor, and thus a thick humus layer; this is composed of faecal pellets of the soil fauna, especially springtails and mites. Litter and fermentation horizons are identifiable in approximately equal thicknesses, and in the case of forms transitional to mull, some incorporation into the mineral soil occurs as well.

Mull forms in freely drained, base-rich soils with good aeration. As these conditions are good for plant life as well, there is a plentiful supply of plant litter, and associated with it a rich soil fauna including earthworms. The organic debris is completely broken down and humidified each year so that none remains from one year to the next. Earthworms, in particular, are responsible for ingesting the plant material and intimately mixing it with the mineral part of the soil. When in the mull form, the humus is finely divided (colloidal) and is intimately associated with the mineral soil, especially the clay with which it forms the *clay-humus* complex. Accumulations of organic matter under wet anaerobic conditions are known as peat. A thin layer of peat is a surface feature of some hydromorphic soils, while thicker occurrences constitute organic soils (pp. 25 and 80).

The chemistry of the soil is concerned largely with the chemical and physical activity of these minute particles of the clay-humus complex. As long ago as 1850 Thomas Way found that soils had the power to retain cations and that natural clays reacted in a similar manner to artificial silicates of lime and aluminium. Subsequent work has shown this piece of research to have been correct, but the mechanism was not understood at the time. Acids, alkalies and their salts when in solution dissociate to a certain degree, that is they form *ions* with positive and negative charges. Water has the same property as well, dissociating into a hydrogen ion H^+ and a hydroxide ion OH^-.

$$NaNO_3 \rightleftharpoons Na^+ + NO_3^-$$
$$H_2O \rightleftharpoons H^+ + OH^-$$

Because of the broken edges of the silicate clay crystals and ionic substitution within their structure, the clay mineral particles have a net negative charge. The clay-humus particle effectively acts as a highly charged anion, which, in colloidal chemistry is known as a *micelle*. Surrounding the micelle and attracted to it by the negative charge are numerous adsorbed cations (Fig. 2.7). These adsorbed cations surrounding the micelle are capable of being exchanged for others. The total amount of exchangeable ions is known as the total exchange capacity of the soil.

2.7 Diagrammatic representation of a clay-humus micelle with the adsorbed ions of hydrogen (H^+), calcium (Ca^{++}) and plant nutrients (K^+) etc.

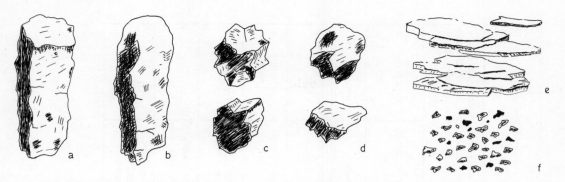

2.8 Soil structures formed by the aggregation of the sand, silt and clay particles: (a) prismatic; (b) columnar; (c) angular blocky; (d) sub-angular blocky; (e) platy; (f) crumb or granular

The process of *leaching* is in effect the continual adsorption of hydrogen or aluminium ions in place of the calcium and other bases. These are displaced by the mass action of hydrogen ions present in carbonated water and the acid breakdown products of plant litter, and of aluminium ions dissolved from the clay minerals themselves in the acid conditions. The soils of the humid temperate regions tend to be saturated with the ions of hydrogen and calcium as well as lesser numbers of magnesium and potassium ions. In the humid tropics aluminium ions play an important part. With increasing dryness of climate and less leaching, calcium and magnesium ions dominate the exchange positions of the clays as in the chernozem soils. Where leaching is minimal and the groundwater rich in soluble salts, including sodium, the exchange positions become dominated with sodium ions. This last condition is found in the semi-desert and desert areas of the world where salt- and sodium-saturated soils limit agricultural production. Physiologically, plant growth is limited by salt in the soil and also by its poor physical conditions. Clays which are saturated with calcium and hydrogen ions are stable and the soil crumbs are flocculated, whereas, with sodium as the predominant ion, the clay particles are dispersed and the soil becomes structureless and difficult to cultivate.

Soil structures

Soil structure is an important physical characteristic of any soil. Structure in a soil is brought about by the individual particles of sand, silt or clay aggregating together in larger units known as *peds*.

Soil structure is encouraged by the incorporation of organic matter, the gums and mucilages formed in the bacterial breakdown of organic matter help to bind the peds together. The peds have been described as the 'architecture' of the soil, and the spaces around them act as channels for the conducting of water through the soil. The spaces between the peds are also important for the soil microfauna which live in them. In fact the volume of the spaces in an organic-rich, medium-textured soil can be as high as 60 per cent in the topsoil, but is usually around 50 per cent. Cultivation reduces the number of the larger pores which are valuable for the movement of air and water through the soil. An average figure of 45 per cent is quoted by Buckman & Brady for nineteen cultivated Georgia soils with 57 per cent pore space in neighbouring uncultivated soils.

Soil structures readily fall into five categories: *structureless*, *platy* structures, *crumb* structures, *blocky* structures and *prismatic* structures. A pedologist will usually describe the type, size and how well-formed the structures are in a soil. The different types of structure are described in Table 2.2 and depicted in Fig. 2.8. Maintenance of soil structure is important for agriculturists the world over. Unless a soil is well-structured, crop yields are depressed, and soils are more liable to erosion. Structure can be weakened by over-cropping, or by the action of heavy machinery passing over the soil.

Soil air and water

The atmosphere penetrates down into the soil along the fissures and pores between the soil

Table 2.2 *Types and classes of soil structures*

Class	TYPE (Shape and arrangement of peds)						
	Platelike with one dimension (the vertical) limited and greatly less than the other two; arranged around a horizontal plane; faces mostly horizontal	Prismlike with two dimensions (the horizontal) limited and considerably less than the vertical; arranged around a vertical line; vertical faces well defined; vertical angular		Blocklike; polyhedronlike, or spheroidal, with three dimensions of the same order of magnitude, arranged around a point			
				Blocklike; blocks or polyhedrons having plane or curved surfaces that are casts of the moulds formed by the faces of the surrounding peds		Spheroids or polyhedrons having plane or curved surfaces which have slight or no accommodation to the faces of surrounding peds	
		Without rounded caps	With rounded caps	Faces flattened; most vertices sharply angular	Mixed rounded and flattened faces with many rounded vertices	Relatively non-porous peds	Porous peds
	Platy	Prismatic	Columnar	(Angular) Blocky[1]	Sub-angular Blocky[2]	Granular	Crumb
Very fine or very thin	Very thin platy; <1 mm.	Very fine prismatic; <10 mm.	Very fine columnar; <10 mm.	Very fine angular blocky; <5 mm.	Very fine sub-angular blocky; <5 mm.	Very fine granular; <1 mm.	Very fine crumb; <1 mm.
Fine or thin	Thin platy; 1 to 2 mm.	Fine prismatic; 10 to 20 mm.	Fine columnar; 10 to 20 mm.	Fine angular blocky; 5 to 10 mm.	Fine sub-angular blocky; 5 to 10 mm.	Fine granular; 1 to 2 mm.	Fine crumb; 1 to 2 mm.
Medium	Medium platy; 2 to 5 mm.	Medium prismatic; 20 to 50 mm.	Medium columnar; 20 to 50 mm.	Medium angular blocky; 10 to 20 mm.	Medium sub-angular blocky; 10 to 20 mm.	Medium granular; 2 to 5 mm.	Medium crumb; 2 to 5 mm.
Coarse or thick	Thick platy; 5 to 10 mm.	Coarse prismatic; 50 to 100 mm.	Coarse columnar; 50 to 100 mm.	Coarse angular blocky; 20 to 50 mm.	Coarse sub-angular blocky; 20 to 50 mm.	Coarse granular; 5 to 10 mm.	
Very coarse or very thick	Very thick platy; >10 mm.	Very coarse prismatic; >100 mm.	Very coarse columnar; >100 mm.	Very coarse angular blocky; >50 mm.	Very coarse sub-angular blocky; >50 mm.	Very coarse granular; >10 mm.	

[1] Sometimes called nut. The word 'angular' in the name can ordinarily be omitted.
[2] Sometimes called nuciform, nut or sub-angular nut. Since the size connotation of these terms is a source of great confusion to many, they are not recommended.
Source: *Field Handbook*, Soil Survey of England and Wales, 1960.

structures. The soil atmosphere is a natural continuation of the atmosphere above the soil, but, although it is similar in some respects, it differs in others. Compared with atmospheric air, soil air is usually saturated with water vapour and is richer in carbon dioxide.

Average composition of soil air
(per cent by volume)

	Soil air	Atmospheric air
Oxygen	20·65	20·97
Carbon dioxide	0·25	0·03
Nitrogen	79·20	79·00

In the figures given above, the soil air has slightly less oxygen and more carbon dioxide, but the amounts of both these gases vary considerably according to the activity of the micro-organisms living in the soil. Additions of leaf litter or organic manure greatly stimulate the bacterial activity which results in a depletion of oxygen in the soil air and an increase in the amount of carbon dioxide. The exchange of oxygen and carbon dioxide with the atmosphere takes place by diffusion, a process which is hindered if the soil pores and fissures are small and limited in number. If the pores and fissures are filled with water, fresh oxygen cannot easily diffuse in, and such oxygen as may be present is soon used, producing anaerobic conditions. It is in these conditions that the growth of most plants is inhibited and the process known as *gleying* is brought about.

The presence of air and water in the soil is almost complementary, for if the soil is saturated with water the air is driven out. In a saturated soil almost all the pore spaces and fissures between the peds are occupied by water. If the soil is allowed to drain so that all the water contained in the larger pores and cavities is removed, the water which is lost is known as *gravitational water*. After about two days following flooding or heavy rain, a freely drained soil has lost its gravitational water and is said to be at *field capacity*. In this state, considerable amounts of water are still held in the finer pores and by capillary attraction. If the soil continues to lose moisture from these reserves of capillary water, the situation is reached where plants cannot obtain enough water to continue transpiration. Wilting then takes place from which the plant does not recover, this is the *permanent wilting point*. This point is not fixed and will vary according to the soil and plant concerned: for example some desert plants can obtain water from the soil against strong capillary forces, whereas a hydrophilic plant soon succumbs to dry conditions.

Further amounts of water can be obtained from soil in the laboratory. It is possible to bring the soil to air-dry conditions, but as this is a rather variable state, depending upon the humidity of the atmosphere, an oven-dry basis is used for most laboratory determinations. Water held at temperatures up to oven-dry (105 °C) is referred to as *hygroscopic water* and is largely unavailable to plants. This water is virtually part of the mineral structure of the soil (Fig. 2.9).

Water movement in soils can take place through the soil pores by saturated flow in which the rain or irrigation water infiltrates at the surface and continues to percolate downwards by gravity and by capillary forces. In unsaturated soils the movement is much restricted, taking place slowly in response to the capillary forces. Because of the variation of size of the pores and also the entrapped air, the capillary rise is not so great as it might be in a true capillary tube. Thus the rise of water from a water-table to the surface is probably limited to about a metre at the most. Water can also move in the vapour phase from a warmed soil layer into a cooler layer where condensation takes place. This form of movement is not so important, except possibly in desert soils, as the movement by gravitation and capillarity.

The soil water will dissolve any soluble constituents which may be present and as such contribute to the *soil solution* which is the medium whereby plants are supplied with nutrients. As has

2.9 Soil water. The width of a film of moisture around a solid soil particle determines the tension at which it is held. Represented in atmospheres, this tension holds water loosely at its outer edges, but as the water film is narrowed by drying, it becomes progressively more tightly held

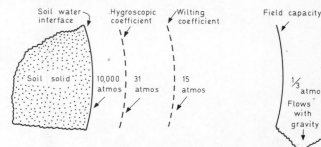

already been observed, inorganic salts dissociate into ions in solution. Many of these ions are attracted and adsorbed to the clay-humus micelles, but an equilibrium is reached so that there is a relationship between the numbers in the exchange positions and the ions still in the soil solution. In the case of the hydrogen ions, their concentration in solution is indicated by the pH scale in which neutrality is pH 7. Values below pH 7 are acid and above pH 7 are alkaline. The average pH range of soils is from below pH 3 to pH 10, but in humid regions the normal range is from pH 5 to pH 7, and in arid regions it is from pH 7 to pH 9. Acid sulphate soils may reach pH 2, whereas at the other end of the scale, alkaline mineral soils may reach a pH value of 10.

Soil pH represents an easily determined feature of a soil, and one which has a general usefulness. Because several plant nutrients become less available to plants at the extremes of pH values and other elements become available in toxic amounts, the pH value is often a guide in the diagnosis of fertility problems. Soil pH values are also used in the classification of soils as can be seen in Chapter 4.

In conclusion it must be pointed out that the soil solution is responsible for the removal of constituents from one horizon, as well as transport and eventually redeposition in a lower horizon. Constituents may be taken into true solution if they are soluble, but often removal takes place linked to other constituents such as the relation of iron and organic molecules.

3 FACTORS AND PROCESSES OF SOIL FORMATION

The consideration of how a soil forms inevitably leads to the question of the environmental controls affecting the manner in which a soil develops. The famous Russian soil scientist Dokuchaiev suggested five soil formers which controlled soil formation. These were: *parent material, climate, age of land, plant and animal organisms, topography.* In other words, the parent material is acted upon by the climate and the organisms, over a period of time, signified by the age of the land. Topography was also considered because this has much to do with water relationships. In more recent years, an American soil scientist, Jenny, considered similar soil-forming factors, and went on to show how they were functionally related in a form of an equation

$$s = f'(\text{Cl, O, R, P. T})$$

in which s, a soil property, is dependent upon (or is a function of) the soil-forming factors climate (Cl), organisms (O), relief (R), parent material (P) and time (T). The usefulness of this approach is that it is possible to take any soil-forming factor and consider its variations against the background of the others, thus examining the effect of that particular one. In view of the importance of these soil-forming factors a brief discussion of their individual roles in soil formation will be considered.

Climate

Climate is a composite concept which includes temperature, humidity, evapotranspiration as well as the type and amount of rainfall, duration of sunshine and many other variables. There is considerable range in world figures: rainfall for example varying from less than 250 mm. per annum to more than 12,500 mm. per annum and annual temperatures which range through 43°C as well as those which range by only 0·5° or 1·0 °C. However the annual figures themselves are not very suitable for a classification system. It is more important to assess the effect of seasonal variation as in the steppes and savanna lands, or the intensity of the rainfall, e.g. as torrential downpours or as a gentle drizzle. These, as well as the local micro-climatic effects, are most important but are unfortunately not often recorded. The rainfall which eventually penetrates into the soil is that which remains after losses by run-off and evaporation direct from the vegetation. Thus the moisture which enters the soil is less than the rainfall. It is further diminished by losses incurred by evapo-transpiration from the soil surface and from moisture taken up by roots and transpired from the leaf surfaces of plants (Fig. 3.1). A number of attempts have been made to reduce the effects of rainfall and evapotranspiration to a single figure. Each has its merits, but lacks a world-wide

3.1 Water entering the soil is not the same as the rainfall; the diagram shows the losses incurred before the water reaches the soil. Some water may be received by movement downslope through the soil

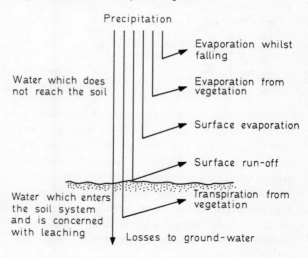

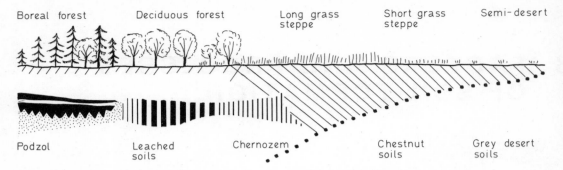

3.2 The amount of rainfall can be broadly correlated with the depth to which calcium carbonate is leached. Thus it is completely removed in the podzol and brown earth and can appear on the surface of a desert soil

significance. Thornthwaite's precipitation : effectiveness index works well in North America, but is less accurate in its assessment of the climate of Europe, to quote but one example. It is possible to show correlations with rainfall for a number of different soil characteristics. One obvious example is the leaching of calcium carbonate from soil; the greater the rainfall the deeper the horizon of calcium carbonate appears in the soil until it is leached right out (Fig. 3.2). Clay type and content in soils is also slightly correlated with rainfall.

Temperature at first sight may not seem to have such an important role to play, but following an idea of Ramann, it is possible to show how it can be important over a long period of time.

	Arctic	Temperate	Tropical
Average soil temperature	10	18	34
Relative dissociation of water	1·7	2·4	4·5
Days weathering	100	200	360
Weathering factor (absolute)	170	480	1620
Weathering factor (relative)	1	2·8	9·5

(From H. Jenny)

In this table the relative dissociation of water is taken as an index of the rate of chemical activity, and this, multiplied by the length of the weathering period, gives the absolute weathering factor. It can be seen that in tropical regions the effectiveness of weathering is almost ten times that of arctic regions and more than three times that of temperate regions. In tropical regions there is the additional fact that weathering has not been interrupted by a change in climate such as the glacial periods of higher latitudes. Therefore, deeper weathering is characteristic of the tropical regions of the world. Up to 15 m. of weathered mantle often occurs though depths are very variable. Correlations with temperature show an increase of temperature occurring with an increase in clay content. The increased rate of chemical activity described above is seen also in the organic decomposition of plant litter which is more rapid in tropical soils.

Organisms

The role of organisms in soil formation is of critical importance for without life there can be no soil formation. Bacteria and fungi are both responsible for the initial breakdown of plant tissue upon and within the soil surface. Various soil arthropods take the breakdown a stage further by eating through the plant remains. Mites (*Arachnidae*) and Springtails (*Colembollae*) are chiefly responsible for this (Fig. 3.3). Incorporation of the organic matter into the mineral soil and the intimate association of the mineral and organic constituents are accomplished by earthworms or by termites.

Although the vegetation is dependent in the last instance upon the climate, it can also function as an independent variable. For example where coniferous plantations have replaced deciduous trees, podzols have formed, and where oak woodlands have developed on chernozems, leaching has been encouraged. Much greater changes occur when forest is cleared for agricultural development. The supply of organic matter is interrupted and changed; plant nutrients originally circulating in the soil-plant system are taken out in the form of crops, and others are added in the form of fertiliser applications. The whole soil environment can be changed by the addition of lime to raise the pH value. The installation of a drainage system changes the hydrological relationships while cultivation mixes horizons previously distinct. The activity of man considerably affects soil development.

Relief

The effect of relief upon climate is well known for its depression of temperature and increase in rainfall. However, associated with upland areas is a greater incidence of cloudiness and hence less solar warming; evapotranspiration is also reduced, leading to conditions favouring the development of thicker organic horizons. Such conditions are particularly prevalent in the northern and western parts of the British Isles. The aspect of slopes can greatly affect the amount of solar warming, so that different soils will form on sun-facing slopes as compared with shady slopes (Fig. 3.4). Amounts of run-off and infiltration will vary also. Where very steep slopes occur, the rate of erosion may mean that a mature soil profile does not have a chance to form; accordingly some soil classifications distinguish a group of mountain soils. Although not so obviously seen in the British environment, it is possible for the transfer of soil constituents to take place downslope. This lateral migration of material brings about the relationship of soils to their position on the landscape. This interrelationship is known as a *catena* (pp. 34 and 69):

Parent material

The parent material is described by Jenny as 'the initial state of the soil system'. It is more generally described as the consolidated or unconsolidated material little affected by the present weathering cycle from which the soil has developed. The more simple definition, that it is 'that which lies beneath the true soil horizons' could be misleading, as many soils have developed from diverse parent materials of differing origins, and not simply from the rock beneath them. For example a soil can develop in a glacial drift or *loess* overlying an unweathered rock which is unaffected by the current processes of soil formation. If the cover of superficial material is thin, then the soil can be developed in both it and the bedrock beneath.

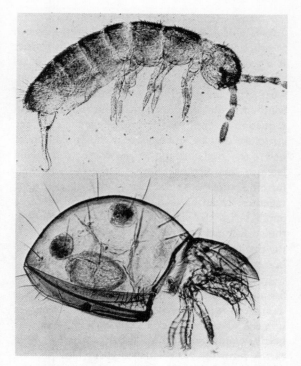

3.3 Soil fauna: examples of a collembola and a mite

3.4 The effect of aspect and relief: (a) soils on shady and sunny slopes (b) normal, receiving and shedding sites

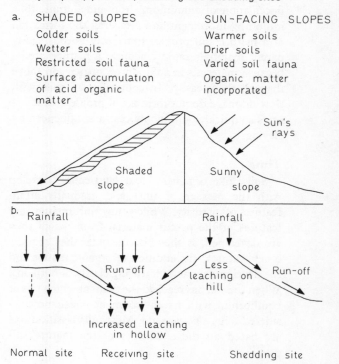

a.   SHADED SLOPES   SUN-FACING SLOPES
     Colder soils   Warmer soils
     Wetter soils   Drier soils
     Restricted soil fauna   Varied soil fauna
     Surface accumulation   Organic matter
     of acid organic   incorporated
     matter

It is useful at this point to distinguish between weathering and soil-forming processes, the former is a geological process, and the latter a pedological process. In addition to physical breakdown, weathering includes the geo-chemical processes of solution, hydrolosis, carbonization, oxidation and reduction as well as the rearrangement of the structure of clay minerals and the formation of new ones. Soil-forming processes are considered later in this chapter. Both weathering and soil-forming processes can carry on separately or together. In an old soil, weathering will occur at depth below the soil, which has been forming at the surface, and in a young soil which is shallow the two processes can operate within the same few inches of soil.

In most environments the balance of soil-forming factors is such that different rocks when weathered will give rise to different soil types because of their mineral composition. This has been demonstrated in Scotland, where brown earths form on basic igneous rocks and podzols form on acid igneous rocks under identical weathering conditions. The differences between soils developed over calcareous and non-calcareous parent materials is recognised in most soil classifications. The presence of calcium causes the flocculation of aluminium, iron and humus, thus inhibiting movement and retarding the formation of a mature soil profile. Extremely porous sandy parent materials rapidly achieve a mature profile because materials in solution or suspension move through them easily. In contrast, clay soils with slow drainage do not form zonal profile characteristics so quickly, and in any case are influenced by gleying.

Time

Soils, like organisms and landscapes, develop with the passage of time and gradually attain features of maturity. While young soils retain many features of the parent material from which they are developed, as they become older they acquire other features: the addition of organic matter and the development and increasing clarity of horizons. When the soil has reached the point at which it is at equilibrium with its environment it can be considered to be a *mature* soil. Most soil classifications are based on the characters of the mature soil profile.

There are several examples of the formation of soils which can be said to have taken place for a definite number of years. Often these examples are related to catastrophic events like the eruption of Krakatoa in 1883, or to the development of soils on landslides and earthflows. The retreat of glaciers has left behind an expanse of parent material upon which soil formation has begun. As records have been kept of the former positions of glacier snouts it is possible to date the beginning of soil formation accurately.

The draining of a lake exposes its floor to the processes of soil formation. The draining of the Ijsselmeer polders has provided much interesting information upon what the Dutch call the *ripening* of soils. The development of series of parallel dunes has again given some indication of the rate at which soils develop. An examination of the soils which are developing upon mining spoil heaps shows sequences of maturity related to the amount of time elapsed since the material was dumped.

Where older soils occur, as in the tropical regions, it is possible to establish a sequence of soil maturity correlated with the morphology of the landscape thus linking together two branches of geographical study. The position of the soil on the landscape gives its age relationships (Fig. 3.5). In this case the time sequence of the geomorphologist is similar to that of the pedologist, though for most soils the period of formation is shorter. In the British Isles it is generally considered that soil formation dates from the end of the Pleistocene Glaciations about 10,000 years ago. Although soils formed in the previous interglacial periods were eroded away, pockets of deeper weathered material in Scotland and South Wales indicate that a former period of soil formation took place in the warmer climate of interglacial times. Changes of climate are known to have occurred since, and following the Iron Age period of British history increased podzolization occurred in southern Britain. Many profiles have features which suggest that the present soil-forming factors have changed slightly. Lower horizons, which bear the imprint of different conditions, have yet to be brought into equilibrium with the present environment.

The processes of soil formation

The processes of soil formation are those which modify the already weathered regolith and eventu-

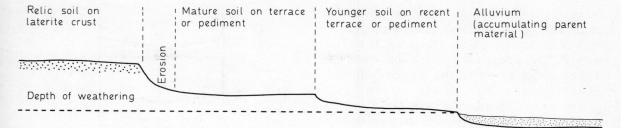

Relic soil on laterite crust

Mature soil on terrace or pediment

Younger soil on recent terrace or pediment

Alluvium (accumulating parent material)

Erosion

Depth of weathering

3.5 Age and position of soil on landscape. Different geomorphic surfaces give rise to a sequence of younger, mature and relic soils in certain parts of the world

ally give it the acquired characteristics which distinguish soil from parent material. Podzolization, calcification, ferrallitization, gleying, salinization, solodization and the formation of peat comprise those processes of soil formation which are generally recognised as controlling the development of the soils of the world. These processes are not mutually exclusive. Gleying and podzolization can be seen to be taking place in the same soil at the same time. Similarly, formation of a laterite is very largely the result of gleying in a tropical environment with specific topographical conditions.

The process of *podzolization* is prevalent in the soils of the cool humid parts of the world and produces soils of the *podzolic group* and the *podzols* in particular. These two groups of soils can be differentiated because in the case of the development of podzols, the processes are more severe and the profile formed is more distinct both in appearance and in its physical and chemical properties. Characteristically, podzols are developed beneath heath or coniferous forest. Podzolic or leached soils are more widely developed under a range of different vegetation covers including deciduous forest, as well as pasture and arable land where podzolization is not so strongly developed.

Dealing with the formation of podzols first, this process involves the development of an extremely acid humus formation known as mor, which results from the accumulation of the litter of acidophilous plants such as heath species and coniferous trees (Fig. 2.6). The rate of decomposition is slow, so that litter accumulates forming fermentation and humus horizons. Rainwater percolating through these superficial horizons becomes strongly acid and acquires water-soluble breakdown products of the plant material. In the strongly acid conditions these breakdown products are capable of forming

complexes with iron in the soil which is then removed as the solutions percolate downwards. In this state iron is said to be *mobilised*. In the same manner, aluminium is also mobilised, leading to a breakdown of clay minerals – largely the alumino-silicates (Fig. 3.6).

There is therefore a tendency for quartz silica to accumulate in the immediate sub-surface, thus

3.6 The process of podzolization

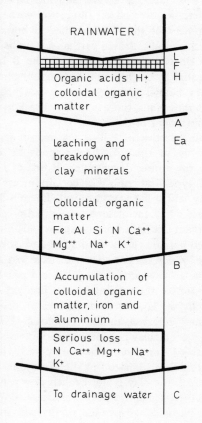

RAINWATER

L
F
H

Organic acids H⁺ colloidal organic matter

A
Ea

Leaching and breakdown of clay minerals

Colloidal organic matter Fe Al Si N Ca⁺⁺ Mg⁺⁺ Na⁺ K⁺

B

Accumulation of colloidal organic matter, iron and aluminium

Serious loss N Ca⁺⁺ Mg⁺⁺ Na⁺ K⁺

To drainage water

C

forming the bleached grey horizon. This is a relative accumulation though, as in the presence of organic matter, some silica is reported to be lost as well. This lighter-coloured horizon which has lost constituents by eluviation is described as an *eluvial* horizon or *albic* horizon (p. 29).

The iron oxides which are mobilised from the surfaces of the mineral soil particles, together with the aluminium and organic matter, are moved in the soil solution down through the profile. There are several different opinions as to the reason for the deposition of material in the B horizon, lower in the profile. These explanations include wetting and drying cycles, changes in pH value, presence of flocculating ions and the breaking down of the iron-humus bonds by ageing or by bacterial attack. The formation of these *illuvial* horizons constitutes the *spodic* horizon of the *7th Approximation* soil classification (p. 29).

A strict interpretation of the *brown earth soils* includes only those soils developed under a moder or mull humus which are strongly leached. These soils lack the eluvial and illuvial horizons characteristic of podzols and are without the development of the clay-enriched B horizons of the leached soils. Brown earths (*sols bruns acides*) lack distinct horizons except that the B horizon may be slightly brighter in colour than the horizons above or below. Although strong leaching takes place in these soils, there is no clay breakdown and little movement of iron.

In the production of a leached or podzolic profile, as in the grey-brown podzolic or *sols lessivés*, the humus development is that of a moder or mull type (Fig. 2.6). The conditions are not so strongly acid as in the podzol soil. There is a richer soil fauna which facilitates a more rapid breakdown of the plant debris. In most leached soils a lighter-coloured horizon can be distinguished below the A horizon, which can be shown to have lost clay by *lessivage*. In this process the clay is washed down the fissures of the soil into the B horizon so that eventually a clay-enriched horizon is produced, sometimes referred to as a *textural B horizon* (Fig. 3.7). The clay content can be increased by as much as two or three times compared with that present in the E horizon of the same soil. Unlike the podzol, the clay is not broken down, and the process is regarded as a purely mechanical washing of clay particles, suspended in the soil solution,

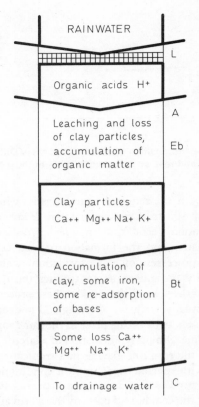

3.7 The process of leaching

into the lower horizons of the soil. When this clay is deposited, it takes up an oriented position parallel to the ped surface upon which it is deposited. In well-developed examples, *clay skins* can be seen with the aid of a hand-lens or even with the naked eye (Figs. 3.8 and 9). In the terms of the recent American classification this horizon would be known as an *argillic* horizon (p. 29).

The process of *calcification* is characteristic of low-rainfall areas in continental interior situations. Leaching is slight and although downward movement does take place, the soluble constituents are not removed from the soil profile. These soils are only wetted to a depth of between 1 m. and 1·5 m. when the moisture begins to re-evaporate (Fig. 3.10). A *calcic* horizon of calcium carbonate accumulates where the impetus of the downward percolating rainwater (or snow melt) is lost. As these soils are relatively unleached, the exchange capacity is dominated with calcium ions, and to a lesser extent with magnesium ions. The presence

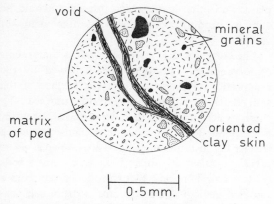

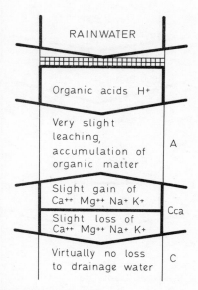

3.8 A thin section of soil as seen through a microscope showing an oriented clay skin lining the walls of a pore

3.10 The process of calcification

of these ions has a stabilising effect upon the colloids and movement in the soil is inhibited.

The type of humus is mull, produced by the natural vegetation of grasses. These grasses have intensive root systems which when dead provide

3.9 Thick clay skin in pore and thin clay skins on ped surface

large amounts of organic matter to the soil. The aerial parts of the plant also return bases to the soil surface. Winter frost and summer drought combine to limit the rate of decomposition so that over many years a very rich and deep A horizon accumulates.

The process of soil formation in the humid tropics has been given many different names by American, Belgian, British and French scientists. The process concerned with changing the parent material into a soil consisting of kaolinite and sesquioxides is known as *ferrallitization*. The use of this word is preferable to the words laterite or latosol whose meaning and use has become confused.

In the humid tropics conditions of abundant rain, high temperatures, and a long uninterrupted period of weathering and soil formation have combined to produce a deep and strongly weathered parent material. The leaf fall from the luxuriant forest and its rapid decay keeps bases in rapid circulation between the soil and vegetation. Conditions are moderately acid in the soil and silica is thought to be removed in preference to the iron and aluminium which gradually accumulate, and oxidise to the sesquioxides in conditions of free drainage (Fig. 3.11).

Rather different conditions obtain where there is a marked dry season as in some of the tropical

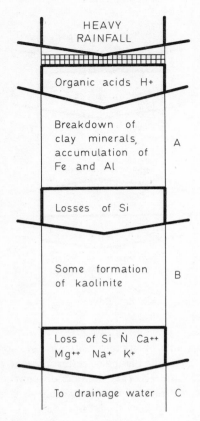

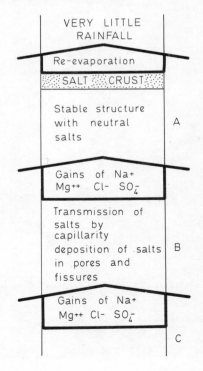

3.11 The process of ferrallitization

plains and other areas which otherwise would be able to produce considerably greater yields of crops.

The enrichment of a soil with salt is the process of *salinization*. This is usually achieved by the evaporation of moisture from the surface of the soil. Salts in solution are drawn upwards by capillary action and are then deposited as the water is evaporated. As a result, these soils develop a surface encrustation of salt and are known as *white alkali* or *solonchak* soils (Fig. 3.12). Such soils possess *salic* horizons. Salt can be derived from a salt-rich geological substratum, or from salt sea-spray blown inland which gradually accumulates in the unleached soils of the arid and semi-arid areas.

When the sodium ion dominates the exchange positions of the clay-humus complex, it becomes dispersed, producing an unstable soil structure. The dispersed humus gives a dark colour and the pH is higher, hence these soils are called *black alkali* or *solonetz* soils. The slight leaching which occurs removes the soluble (neutral) salts. Thus the clay-humus complex is dominated with

3.12 The process of salinization

grasslands. The soil is strongly dried for part of the year and in the rainy season is strongly leached. These alternations may well increase the rate of chemical activity and movement of soil constituents. Many of the soils with iron crusts have formed in conditions such as these.

The strongly leached conditions of the parent material of many tropical soils in positions with free drainage leave only the possibility of kaolinite forming by recombination of elements in the soil solution. However, in receiving sites where the supply of bases is plentiful, montmorillonite clays are more likely to form and be the dominant clay mineral.

In arid climates the rainfall is irregular and insufficient to remove soluble salts from the soil, although in semi-arid areas there is a redistribution of the salts into the lower parts of the landscape. The occurrence of soils affected by salt is associated with the soils which have imperfect or poor natural drainage. Frequently these areas are the alluvial sodium ions giving the soil a *natric* horizon. The process of *solonization* or removal of sodium ions from the clay-humus complex, results in a range of soils from the solonetz through an intermediate form, the *solodized solonetz*, to the *solod* which is largely leached of metal cations and is dominated with hydrogen ions, resulting in an acid soil (Fig. 3.13).

The presence of water in a soil for long periods brings about anaerobic conditions as has been explained in Chapter 2. As this can happen in almost any type of zonal soil, the process of gleying produces features which enable the pedologist to group these soils together as *hydromorphic* soils (Chapter 10). These poorly drained or hydromorphic soils frequently occur in the lower parts of the landscape, and are often developed from parent materials similar to those of adjacent freely drained soils.

Two basic types of gley soil are recognised. Where an impervious horizon occurs within the

soil profile there results a *surface-water gley*, while water rising in the soil from an impervious horizon beneath the soil, produces a *ground-water gley*. It is usual to find ground-water gley soils in areas which receive an inflow of surface and ground-water. In this case, the soil must be permeable if water rising between the grains or peds is to saturate the lower soil horizons. With anaerobic conditions the decomposition of plant residues on the surface is slow and peat frequently accumulates on soils with a high level of ground-water. The colour of these soils may frequently be an unrelieved grey of ferrous iron, or, if the soil dries occasionally, there may be a few mottles of rust-coloured ferric oxide. Surface-water gley soils can occur almost anywhere on the landscape where an impermeable horizon such as a Bt horizon prevents the downward passage of water through the profile. Localised gleying results, particularly on the faces and pores of the peds, which become grey coloured. The ped interiors, in contrast, retain the brighter colours of ferric iron, giving an overall impression of mottling when revealed in a section. This soil closely resembles the *pseudogley* described by European authors.

Finally, conditions of very poor soil drainage encourage *peat formation* in association with the gley soils present. Basically there are two types of peat, the moor peat which is acid, and the fen peat which is neutral or mildly alkaline. Further subdivision can be made upon botanical composition, structure and degree of decomposition. The characteristics of these two basic forms result from the way in which they have developed. The acid moor peat develops on upland areas where high rainfall results in the leaching of all bases so that acid conditions set in and there is very slow decomposition of the plant debris. In certain areas deep peat has accumulated to depths of 3–5 m. These are raised bogs. The mosses which grow upon them are sustained from the nutrient supplies of the rainwater. Raised bogs also occur in lowland sites where an acid peat overlies a peat formed in a declivity of the landscape. The true fen peats develop in waters liberally supplied with bases and are neutral or mildly alkaline in reaction. When adequately drained, the soils developed upon such deposits are usually very rich with high land values as in the fens of eastern England.

3.13 The process of solodization

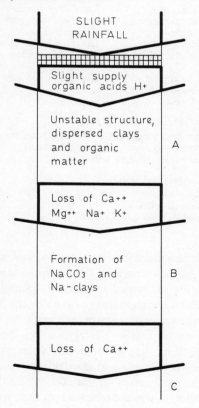

SLIGHT RAINFALL

Slight supply organic acids H+

Unstable structure, dispersed clays and organic matter — A

Loss of Ca++ Mg++ Na+ K+

Formation of Na CO₃ and Na - clays — B

Loss of Ca++

C

4 SOIL CLASSIFICATION

Classification is only a contrivance to order a subject, perhaps to understand and remember its content more easily, certainly to show how one part is related to the whole. A classification serves to make a basis from which further enquiry can proceed. As far as soils are concerned, classification is still developing, and numerous attempts have been made from different points of view to group soils into natural classes. However, a soil which in one place is extensive, elsewhere may only occupy a small and insignificant area, or may not be represented at all. It is in some ways unfortunate that all soil classifications are not marked 'provisional', because as more knowledge of the soil is collected, new groupings are deemed necessary and older ones may be redefined.

Several recent attempts to classify soils are based upon the intrinsic properties of the soils, though in the past this has not always been the case. Many classifications have been based upon an interpretation and comparison of mature soils with attempts at grouping in a logical, natural system of categories. A *mature* soil can be described as one which is in dynamic equilibrium with its environment. A *young* soil is still in the process of adjustment to the current situation. Change of climate, cultivation, clearance of vegetation, lowering of ground-water, removal of salts can all result in readjustments being made in the soil with consequent modifications to the nature and appearance of the horizons present.

It is generally accepted that the impetus of present-day soil science began with V. V. Dokuchaiev in U.S.S.R. and E. W. Hilgard in U.S.A. Dokuchaiev is credited with producing the first natural soil classification, a classification based on observable features in the soil itself. He recognised that the soil was an independent natural body which could be studied by field and laboratory methods, based upon the interpretation of the morphology of the profile. It was observed by Dokuchaiev and his collaborators that many soil types had a definite geographical location, associated with definite climatic regions and vegetation types. This led to the development of the idea of zonal soils which is the main feature of the final classification produced by Dokuchaiev in 1900, and which still influences current world soil maps (Table 4.1).

Table 4.1 *Classification of world soils by V. V. Dokuchaiev (1900)*

Zones	Soil types
Class A. Normal, otherwise dry land vegetative or zonal soils	
I Boreal	Tundra (dark brown) soils
II Taiga	Light grey podzolized soils
III Forest-steppe	Grey and dark grey soils
IV Steppe	Chernozem
V Desert-steppe	Chestnut and brown soils
VI Aerial or desert-zone	Aerial soils, yellow soils, white soils
VII Sub-tropical and zone of tropical forests	Laterite or red soils
Class B. Transitional soils	
VIII Dry land moor soils or moor-meadow soils	
IX Carbonate-containing soils (rendzina)	
X Secondary alkaline soils	
Class C. Abnormal soils	
XI Moor soils	
XII Alluvial soils	
XIII Aeolian soils	

In Britain much interest in the soil was expressed by the various writers of the *General Views of the Agriculture* of particular counties, commissioned at the end of the eighteenth century by the Board of Agriculture, though none of their work was systematic in its content or description. In the period from 1900 to 1914 a number of attempts at soil mapping were made and close relation was found with the drift maps of the Geological Survey. Fortunately it was appreciated that the correlation

Table 4.2   *Classification of world soils by G. W. Robinson (1947)*

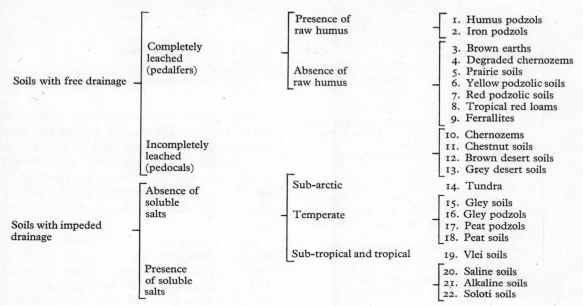

Soils with free drainage	Completely leached (pedalfers)	Presence of raw humus	1. Humus podzols 2. Iron podzols
		Absence of raw humus	3. Brown earths 4. Degraded chernozems 5. Prairie soils 6. Yellow podzolic soils 7. Red podzolic soils 8. Tropical red loams 9. Ferrallites
	Incompletely leached (pedocals)		10. Chernozems 11. Chestnut soils 12. Brown desert soils 13. Grey desert soils
Soils with impeded drainage	Absence of soluble salts	Sub-arctic	14. Tundra
		Temperate	15. Gley soils 16. Gley podzols 17. Peat podzols 18. Peat soils
		Sub-tropical and tropical	19. Vlei soils
	Presence of soluble salts		20. Saline soils 21. Alkaline soils 22. Soloti soils

was not perfect, and in the years following the 1914–18 war visits by American soil surveyors, together with the appointment of soil survey assistants to the regional agricultural chemists led to the adoption of the American system of soil series, based upon the characters of the soil profile.

Much of the early work in soil survey in Britain was undertaken by Professor G. W. Robinson at Bangor. He produced a classification of world soils, but unfortunately did not make a classification of British soils, although he reviewed the present state of the knowledge of British soils at the time he wrote (Table 4.2).

This classification reflects the approach of pedologists to the classification of world soils before and immediately after the 1939–45 war. It is based upon mature profiles and it includes, drainage, degree of leaching, and type of humus in its criteria. For these it draws upon the experience of Marbut (1927) and subsequent modifications of his classification by the American Soil Survey. Robinson also foreshadowed ideas later expressed by Kubiena.

Perhaps the most widely quoted and influential system of classification in Europe is that of Kubiena (1953). He groups profiles by their horizon sequence as designated by the horizon

index letters, A, B, C, etc. while the type of humus plays an important role in the second stage of his classification. With five basic groups, this classification has the great merit of simplicity and gives a good overall view of profile morphology in soil classification. After an initial division on horizon sequence, a further sub-division is made upon water relationships, e.g. below water level, poorly drained and freely drained soils (Table 4.3).

Table 4.3   *General grouping of soils by their profile by W. L. Kubiena (1953)*

1. (A)C-soils
    With soil life, but without macroscopically distinguishable humus layers and only with an upper layer colonised by organisms (with or without a plant root layer)
    Sub-aqueous: underwater raw soils (e.g. deep-sea clays, coral reefs)
    Semi-terrestrial: raw gley soils
    Terrestrial: raw soils, arctic raw soils, desert soils

2. AC-soils
    With distinct humus horizon, but without B horizons
    Sub-aqueous: underwater humus soils (e.g. dy, gytta, sapropel, reed peat)
    Semi-terrestrial: humus-gley soils (e.g. anmoors, gleyed grey warp soils, mull gley soils)
    Terrestrial: rendzina- and ranker-like soils (e.g. rendzinas, chernozems, rankers, para-chernozems)

3. A(B)C-soils

With pronounced B horizons which however are not real eluvial horizons built up by peptizable substances, but whose origin, in the first place, is due to deep-reaching weathering with sufficient aeration and oxidation

Terrestrial: brown earth and red earth-like soils (e.g. brown earths, brown and red loams, red earths, *terra rossa*)

4. ABC-soils

With B horizons which are at the same time developed illuvial horizons, i.e. having a strong enrichment of peptizable substances

Terrestrial: bleached soils (e.g. podzols, bleached brown and red loams, soloti)

5. B/ABC-soils

With strong enrichment of illuvial substances transported to the surface layer in peptized state by intensive capillary rise and irreversible precipitation

Terrestrial: rind and surface crust soils

The great trouble with most soil classifications is that they are produced for semi-natural and natural soils, whereas much of the landscape has been affected by the work of mankind. Certainly, it is possible to have different phases of the same soil series under natural conditions and under agricultural conditions, but most soil classifications especially at a national level are based upon the natural profile, developed under woodland or heath.

There are some properties of a soil which do not change materially with agricultural practice, and of these the texture is the most readily observed feature. Thus, a classification can be built up quite objectively upon soils with uniform textures U, soils with textures which gradually change from the surface to the parent material G, or soils which abruptly change in texture somewhere in the soil profile D. This system has been the basis of a new classification and map of Australian soils by K. Northcote (1960) (Table 4.4).

Table 4.4 *Classification of Australian soils by K. Northcote (1960)*

I Soils with uniform texture profiles U
   Uc coarse textures
   Um medium textures
   Uf fine textures
   Ug fine textures showing cracking

II Soils with gradational upland profiles G
   Gc soils calcareous throughout profile
   Gn soils not calcareous throughout profile

III Soils with contrasting (Duplex) texture profiles D
   Dr red clayey sub-soils
   Db brown clayey sub-soils
   Dy yellow clayey sub-soils
   Dd dark clayey sub-soils
   Dg gley clayey sub-soils

IV Soils with organic profiles O
   No further classification

Other methods of soil classification are aimed at specific purposes. For example, in semi-arid areas where much work has been done on alluvial soils for irrigation projects, the presence or absence of salt is an important criterion in the classification. Most classifications include the acidity or alkalinity (pH value) of soils as a criterion at some stage.

As more knowledge has become available, it has been appreciated that it is sometimes difficult to classify soils in a rigid system because in nature they form a continuum with many gradations. To overcome some of the difficulties the American Soil Survey staff produced a new and comprehensive system of soil classification, known as the *7th Approximation* (1960). This avoids all the old colour names and folk names for soils, and new terms have been evolved to describe soils by diagnostic horizons. An *epipedon* is a surface horizon which is darkened by organic matter, and which includes the eluvial horizons. There are six of these horizons, briefly described as follows:

Mollic epipedon. A dark coloured, thick surface horizon, and with over 50 per cent of the exchange capacity saturated by base cations.

Anthropic epipedon. Similar to a mollic epipedon but with a high amount of phosphate accumulated by long-continued farming.

Umbric epipedon. A dark surface horizon less than 50 per cent of the exchange capacity saturated by base cations.

Plaggen epipedon. A man-made surface horizon more than 50 cm. thick with characteristics that depend upon the original soil from which it was derived.

Histic epipedon. A thin surface horizon, saturated with water for part of the year, and with a large amount of organic carbon.

Ochric epipedon. Epipedons which are too light in colour, too low in organic carbon, or too thin to belong to the above.

Sub-surface diagnostic horizons, of which there are thirteen, occur below these surface horizons. The following brief descriptions summarise the detailed explanations given.

Argillic horizon. An illuvial horizon in which clays have accumulated to a significant extent.

Agric horizon. A compact horizon formed by cultivation which has been enriched by clay and/or humus.

Natric horizon. An argillic horizon with columnar structure and more than 15 per cent saturated with exchangeable sodium ions.

Spodic horizon. A horizon with an accumulation of free sesquioxides and/or organic carbon but not with equivalent amounts of crystalline clay.

Cambic horizon. A changed or altered horizon, including structure formation, liberation of free iron oxides, clay formation or the obliteration of the original structure of the parent material.

Oxic horizon. A horizon with a very low content of weatherable minerals, in which the clay is composed of kaolinite and sesquioxides, having a low cation exchange capacity and poorly dispersible in water.

Calcic horizon. Enriched horizon with calcium carbonate in the form of secondary concretions, more than 15 cm. thick.

Gypsic horizon. Enriched horizon with calcium sulphate, more than 15 cm. thick.

Salic horizon. Enriched horizon with salts more soluble than gypsum, more than 15 cm. thick.

Albic horizon. A horizon from which clay and free iron oxides have been removed, so the colour is determined by the colour of sand and silt and not the coatings on these particles.

Three other horizons are defined which include the indurated horizons of certain soils. These are: the *duripan*, a horizon cemented by silica or aluminium silicate; the *fragipan*, a loamy sub-surface horizon with platy structure and high bulk density, brittle when wet and hard when dry; lastly, *plinthite*, rich in sesquioxides, highly weathered and poor in humus, which irreversibly hardens into crusts and irregular aggregates.

The diagnostic horizons combine together in the *pedon*, a three-dimensional unit of soil with a minimum area of between 1 and 10 sq. m. These are related through *Great Groups* and *Sub-orders* to one of ten *Soil Orders*. The orders and sub-orders are given below with the approximate equivalents in the zonal climatic type of classification (Table 4.5).

Table 4.5 *American Soil Survey staff classification 7th Approximation (1960)*

1. Entisols. Weakly developed (generally azonal) soils

– with features of gleying	Aquents	1.1
– with strong artificial disturbance	Arents	1.2
– on alluvial deposits	Fluvents	1.3
– with sand or loamy sand texture	Psamments	1.4
– other Entisols (e.g. lithosols, some regosols)	Orthents	1.5

2. Vertisols. Cracking clay soils

– usually moist	Uderts	2.1
– dry for short periods	Usterts	2.2
– dry for a long period	Xererts	2.3
– usually dry	Torrerts	2.4

3. Inceptisols. Moderately developed soils, not in other orders

– with features of gleying	Aquepts	3.1
– on volcanic ash	Andepts	3.2
– in a tropical climate	Tropepts	3.3
– with an umbric epipedon	Umbrepts	3.4
– other Inceptisols (e.g. most brown earths)	Ochrepts	3.5

4. Aridisols. Semi-desert and desert soils

– with an argillic horizon	Argids	4.1
– other soils of dry areas (e.g. grey desert soils)	Orthids	4.2

5. Mollisols. Soils of high base status with a dark A horizon

– with albic and argillic horizons	Albolls	5.1
– with features of gleying	Aquolls	5.2
– on highly calcareous parent materials	Rendolls	5.3
– others in cold climates	Borolls	5.4
– others in humid climates	Udolls	5.5
– others in sub-humid climates	Ustolls	5.6
– others in sub-arid climates	Xerolls	5.7

6. Spodosols. Soils with a spodic horizon (e.g. podzols)

– with features of gleying	Aquods	6.1
– with little humus in spodic horizon	Ferrods	6.2
– with little iron in spodic horizon	Humods	6.3
– with iron and humus	Orthods	6.4

7. Alfisols. Soils with an argillic horizon and moderate to high base content

– with features of gleying	Aqualfs	7.1
– others in cold climates	Boralfs	7.2
– others in humid climates	Udalfs	7.3
– others in sub-humid climates	Ustalfs	7.4
– others in sub-arid climates	Xeralfs	7.5

8. Ultisols. Soils with an argillic horizon and low base content

– with features of gleying	Aquults	8.1
– with a humose A horizon	Humults	8.2
– in humid climates	Udults	8.3
– others in sub-humid climates	Ustults	8.4
– others in sub-arid climates	Xerults	8.5

9. Oxisols. Soils with an oxic horizon or with plinthite near the surface

– with features of gleying	Aquox	9.1
– with a humose A horizon	Humox	9.2
– others in humid climates	Orthox	9.3
– others in drier climates	Ustox	9.4

10. Histosols. Organic soils (suborders not finalised)

As can be seen from the above table, the names of the soil sub-orders are compounded from certain formative elements derived from the name of the soil order. Prefixes such as acq-characters associated with wetness, ud-characters associated with humid climates and ust-characters associated with

Table 4.6 *Classification of the major groups of mineral soils by extent of leaching (Hallsworth, 1965)*

Group	Normal soils		Ground-water soils
	Coarse-texture	Fine-texture	
0. Soils showing no or only rudimentary differentiation	Lithosols Aeolian regosols (dune sands) Fluvial regosols		Alluvial soils
1. Soils effectively unleached containing soluble salts, mainly sodium chloride			Solonchak
2. Slightly leached soils (2nd stage of leaching) dominated by sodium ions, often containing gypsum	Sief dune soils Solonetz Solodized solonetz Solodic	Stony gilgai Serozem	Salty alkali Solonchak
3. Moderately leached soils (3rd stage of leaching) dominated by Ca(Mg) and containing secondary carbonate	Red-brown earth Solonized soils	Chernozem Chestnut	Gleyed serozem Gleyed chernozem
4. Moderately leached soils (4th stage of leaching) dominated by Ca(H) and without secondary carbonate	Brown earths Brown soils of light texture Brown limestone soils Sols lessivés	Prairie soils Brown earths Chocolate soils Terra rossa Rendzina	Meadow soils or Wiesenboden Gleyed brown earths
5. Strongly leached soils (5th stage of leaching) dominated by H Al (Fe)	Podzols Grey-brown podzolic Red-yellow podzolic	Krasnozems Red earth	Gley podzols Laterite

dry climates, are added to suggest the properties of the sub-order. Further adjectival prefixes give rise to the great group names. Although the introduction of the American system, has stimulated much discussion, and many references have been made to it, it has not been adopted in Britain or Europe by the soil survey organisations up to the present.

It is unfortunate that many soil classifications have been based upon the presence or absence of certain types of soil humus, for this can easily be changed by clearance or cultivation. Transient features are not the best ones for criteria in a classification. It has already been seen that texture forms a good basis for classification, being one of the more stable features of the soil. Other of the more persistent features are the nature and quantity of the clay minerals present, the amount and type of the exchangeable cations, and the degree of leaching. All of these have been incorporated into a scheme of classification proposed by Hallsworth which seems to have much to offer in a pedological approach to world soils (Table 4.6).

It is the intention to present in this book a review of the main soil types of the world. Because of the complexity at local level, a consideration of soils on a world basis has many disadvantages and inevitably vague generalisations are made. However it is important to realise the many correlations which can be drawn between the different factors of the environment which together produce the soil cover. In fact the soil can be considered to be the product of all these factors or in other words to form a zonal index which signifies a degree of geographical uniformity in the area covered by a particular soil type. The placing of soils in the different environmental zones is following the tradition of Russian soil scientists who, with their opportunity to study soils over continental areas, have one of the best overall views of world soils. As the climate is probably of fundamental importance, the following discussion of the *zonal soils* of the world is based on a broad climatic/latitudinal scheme (Table 4.7).

As these zones cover large and diverse climates mention is made of the climatic conditions where the typical soils are found. Because vegetation responds to the influence of climate and soil, at world level of mapping there is a fairly close correlation between the occurrence of world plant formations and world soils. However, the world plant formation distribution maps, as in the case of soil maps, mask a great variation at local level where parent material and differing hydrological conditions can modify the vegetation and soil type. The presentation here of the associated *intrazonal* and *azonal* soils representative of the different bio-climatic zones is an attempt to overcome the disadvantage of suggesting any one soil for any one bio-climatic zone. The soil properties and distribution will be discussed in the text and the illustrations show relations to landforms in idealised landscapes.

From this chapter it is evident that there are many different ways of classifying soils, and that each system has its merits. However, amongst soil scientists classification is one of the most hotly debated subjects. As Leeper puts it in a discussion of soil classification, 'when scientists discuss methods of analysing a solution for traces of phosphate they are practical, reasonable and unemotional. When the same men discuss the classification of soils these virtues are likely to evaporate'. The problems of soil classification are bound up with local and national prestige, for decisions taken at a local level often cannot be accommodated at a national or international level.

Table 4.7 *A simplified zonal arrangement of soils*

Latitude zone	Bio-climatic zone	Zonal soils
1. Soils of the high latitudes	Tundra	Arctic brown soil
2. Soils of the mid-latitudes, cool climates	Northern forest zone	Podzols Brown earths Leached soils Grey soils
3. Soils of the mid-latitudes, warm climates	Mixed forest zone	Red and brown Mediterranean soils Cinnamon soils Red and yellow podzolic soils
	Steppe zone	Chernozems Chestnut soils
	Semi-desert and desert zone	Grey desert soils Desert soils
4. Soils of the low latitudes	Tropical rain forest and deciduous forest savanna zone	Ferrallitic soils Ferrisols Ferruginous soils Vertisols Laterite

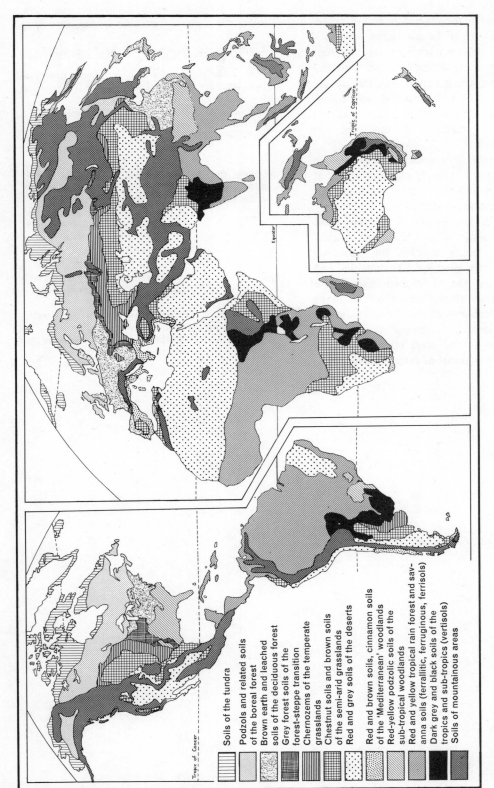

Soils of the tundra

Podzols and related soils
of the boreal forest

Brown earth and leached
soils of the deciduous forest

Grey forest soils of the
forest-steppe transition

Chernozems of the temperate
grasslands

Chestnut soils and brown soils
of the semi-arid grasslands

Red and grey soils of the deserts

Red and brown soils, cinnamon soils
of the 'Mediterranean' woodlands

Red-yellow podzolic soils of the
sub-tropical woodlands

Red and yellow tropical rain forest and savanna soils (ferrallitic, ferruginous, ferrisols)

Dark grey and black soils of the
tropics and sub-tropics (vertisols)

Soils of mountainous areas

5.1 Simplified world soil map

5 SOIL MAPPING

As a natural resource, soil and its distribution is of national interest. The more that is known about its distribution and formation, the more it will be used profitably for the individual and the community. Many countries of the world, including Britain, now possess a soil survey organisation which is charged with finding out the distribution, genesis and properties of soils. From this basic work stems a range of applied uses for pedological knowledge (Chapter 11). Agriculture is the most obvious in that a greater knowledge of the soil will help with decisions about crops, cultivations and fertiliser treatments. Where soils have been maltreated the most effective methods of restoration can be explored and used, as has been the case in the opencast workings for coal and iron in many countries. Natural disasters such as the flooding of arable land by salt water which occurred on the east coast of Britain in 1953, can be combated more effectively if something is known about the nature and distribution

of the soils affected. Engineering scientists are interested in the physical properties of the soil for construction of roads, buildings and embankments. In the developing countries change from an agricultural system of low productivity to a more advanced form necessitates advance knowledge of the soils, particularly where the installation of expensive irrigation projects is being considered.

World soil maps are based essentially on the prevailing environmental conditions, therefore they strongly reflect the patterns seen in the climate and vegetation maps. If the general conditions of soil formation are similar in two given areas, then a specific soil will be produced in both areas. Thus these maps do not always show the actual soils, but only the most probable zonal soil which might occur. These maps show tundra soils, podzols, chernozems, desert soils, etc. (Fig. 5.1). For example, most world soil maps generalise the British Isles into a northwest zone

5.2 In a soil catena a number of different soils are linked together by their relationship to each other on the landscape

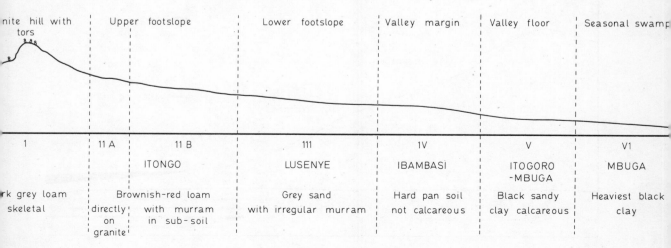

nite hill with tors	Upper footslope		Lower footslope	Valley margin	Valley floor	Seasonal swamp
1	11 A	11 B	111	1V	V	V1
		ITONGO	LUSENYE	IBAMBASI	ITOGORO -MBUGA	MBUGA
rk grey loam skeletal	directly on granite	Brownish-red loam with murram in sub-soil	Grey sand with irregular murram	Hard pan soil not calcareous	Black sandy clay calcareous	Heaviest black clay

of podzols, and a southeast zone of brown earths. This gross over-simplification emerges when it is realised that some of the best developed podzols actually occur in the south and east of England. In producing a world soil map, the process of mapping is carried out at an empirical level, using the climate. Of necessity the scale of such maps prohibits the inclusion of local detail. World maps convenient for reproduction in atlases are usually at scales of about 1:80,000,000 at which only the broadest generalisations can be shown. For general purposes of geographical interpretation it is desirable to have maps of a country or region which show a moderate amount of detail. This becomes possible at scales of 1:5,000,000 and larger. Two groupings of soils are used for these maps, the *catena* and the *soil association*.

The catena was used by Milne for mapping soils in East Africa, where he found a regularly occurring relationship of the soils with topographic features. The type of relationship can be seen from Fig. 5.2. It has been found to be most useful in mapping large areas of similar parent material within a uniform climatic regime.

In a similar manner, the Scottish soil scientists have grouped topographically related soils developed on one geological parent material into a *soil association* (Fig. 5.3). In this case, a catena of soils, based on their inherent drainage properties is made up of a number of soil series. Hence the Ettrick Association derived from Silurian greywackes and shales has six component series in the Jedburgh and Morebattle district:

Freely drained	Linhope series	brown earth
	Dod series	peaty podzol
Poorly drained	Ettrick series	non-calcareous gley
	Alemoor series	peaty gley
Very poorly drained	Peden series	non-calcareous gley
	Hardlee series	peaty gley

Because for reconnaissance work it is convenient to describe the soils of an area collectively, so a slightly different interpretation of the term association is used. In describing the soils which occur on a geographical tract a number of different soils are *associated* in a landscape, even though they may have completely different processes operating in their formation. In the case of the Carboniferous Limestone of central Derbyshire, the associated soils are the Nordrach series, a brown earth; Marian series, a rendzina; and the Ivet series, a brown earth with gleying. These groupings of soils are essentially a regional grouping, bounded by geomorphic elements of the landscape with similar geology, geomorphology, soils and vegetation. Such a mapping unit is called a *land system* by Australian and British soil scientists (Fig. 5.4).

5.3 A drainage association on a permeable parent material

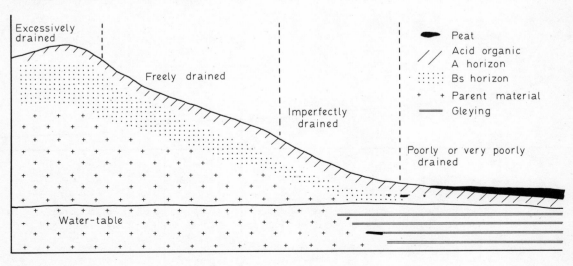

Excessively drained

Freely drained

Imperfectly drained

Poorly or very poorly drained

Peat
Acid organic A horizon
Bs horizon
Parent material
Gleying

Water-table

(3) NAPPERBY LAND SYSTEM (1000 SQ. MILES)

Granite hills and plains with lower rugged country in a strip from Aileron to west of Mt. Doreen homestead.

Geology.—Massive granite and gneiss, some schist. Pre-Cambrian age, Arunta block, Mt. Doreen–Reynolds Range; Lower Proterozoic, Warramunga geosyncline.

Geomorphology.—Erosional weathered land surface: hills up to 500 ft high and plains with branching shallow valleys; less extensive rugged ridges with relief up to 50 ft, and a dense rectangular pattern of narrow steep-sided valleys.

Water Resources.—Isolated alluvial or fracture aquifers may yield supplies of ground water. There are areas suitable for surface catchments.

Climate.—Nearest comparable climatic station is Tea Tree Well.

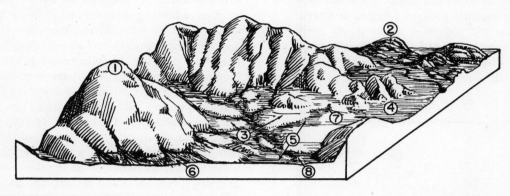

Unit	Area	Land Form	Soil*	Plant Community
1	Large	Granite hills: tors and domes up to 500 ft high; bare rock summits, and rectilinear boulder-covered hill slopes, 40–60%, with minor gullies; short colluvial aprons, 5–10%	Outcrop with pockets of shallow, gritty or stony soils	Sparse shrubs and low trees over sparse forbs and grasses, *Triodia spicata*, or *Plectrachne pungens* (spinifex)
2	Medium	Closely-set gneiss ridges and quartz reefs: up to 50 ft high; short rocky slopes, 10–35%; narrow intervening valleys		
3	Medium	Interfluves: up to 20 ft high and ⅓ mile wide; flattish or convex crests, and concave marginal slopes attaining 2%	Mainly red earths (4*a*), locally red clayey sands (3*a*), and texture-contrast soils (7*a*), stony soils near hills	Sparse low trees over short grasses and forbs or *Eragrostis eriopoda* (woolly-butt)
4	Medium	Erosional plains: up to 1 mile in extent, slopes generally less than 1%		
5	Small	Drainage floors: 200–400 yd wide, longitudinal gradients about 1 in 200	Mainly texture-contrast soils (7*e*), locally alluvial soils (1*a*) and red earths	*Eremophila* spp.—*Hakea leucoptera* over short grasses and forbs; minor *Kochia aphylla* (cotton-bush)
6	Small	Alluvial fans: ill-defined distributary drainage; gradients above 1 in 200	Alluvial brown sands (1*a*) and red clayey sands (3*d*)	Sparse low trees over short grasses and forbs or *Aristida browniana* (kerosene grass)
7	Small	Rounded drainage heads: up to 200 yd wide and 5 ft deep on the flanks of unit 3	Red earths	Dense *A. aneura* (mulga) over short grasses and forbs
8	Very small	Channels: up to 50 yd wide and 5 ft deep and braiding locally	Bed-loads mainly coarse grit	*E. camaldulensis* (red gum) – *A. estrophiolata* (ironwood) over *Chloris acicularis* (curly windmill grass)

* The numbers in parentheses in this column refer to soil groupings in Part VIII.

5.4 Reconnaissance mapping of soils can be done by the 'Land System' method

Detailed soil surveys at scales of 1:63,360 and larger necessitate the examination of the soil at frequent intervals over the landscape so that the profile characteristics are known and the boundaries between them drawn. The normal procedure is first to make a reconnaissance of the area to find the range of soils present. Once the framework is established, it is possible to begin mapping individual soils as mapping units. These mapping units can then be classified into the named soil series or soil complexes which occur in the area being mapped. In Britain, the soil series is defined by G. W. Robinson as follows: 'soils with similar profiles derived from similar material under similar conditions of development are conveniently grouped together as a series'.

Where the pattern of different soil series becomes too complicated and it is deemed uneconomic to map out tiny individual areas of each soil the whole area can be mapped as a *soil complex*.

5.5 Example of detailed soil mapping

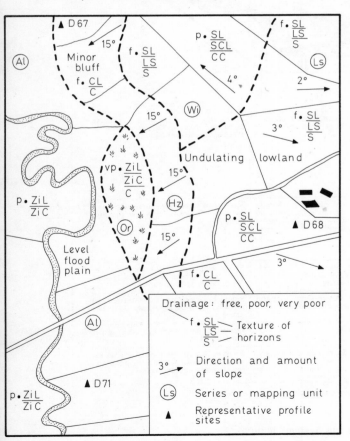

For example, on steep slopes, the variability of parent material and drainage can be considerable, causing a most complicated soil pattern.

Several approaches to detailed mapping are possible, the first method is to observe the soil at fixed points by a grid pattern surveyed over the landscape. This is a method that can be used at a very detailed scale, for example when a field is being surveyed before an agricultural experiment is laid down. In this case the soil would be examined at a close interval of 25 or 50 m., the observations made and the boundaries drawn by interpolation. A similar method of survey is adopted in the case of well-wooded country when it is impossible to locate accurately the position of observations by other means. In West Africa, survey lines are followed and the soil described and sampled at intervals of 200 m. along survey lines 800 m. apart. Supplementary information is obtained from lines at 100 m. interval. This method has been used in the production of soil maps at 4 in. to 1 mile in West Africa (Vine); and the information gained is used to lay out plantations. The production of semi-detailed maps such as these often entails a certain amount of basic survey, as well as observations on the hydrology, vegetation and present land use.

The second method of ground survey is a more interpretative approach in which the morphology of the landscape is used to elucidate the soil pattern. Depending upon the 'lie of the land' the soil is inspected with spade or screw auger and the resulting information marked upon a field map. In Britain the 1:10,560 maps of the Ordnance Survey are the most convenient for routine mapping at a detailed scale. The colour, texture, drainage characteristics and any other notable features are recorded (Fig. 5.5). The boundary between different soil-mapping units is frequently found to follow pronounced morphological breaks of slope in the landscape which aids their location on the map. As the morphology often gives a clue to the origin of the parent material upon which the soil is developed, so the particular soil associated with the deposit of geological nature has a boundary which coincides with its areal distribution.

A third method is to follow a boundary, crossing and recrossing it until its position is accurately known. This is a time-consuming activity and the results are not materially better than those achieved by interpretative methods of mapping. It is a method which has to be adopted when other features such as vegetation and morphology of the land cannot be interpreted. Fortunately, there are very few areas of the world which are featureless, so the surveyor usually has some other data to guide him in his map compilation.

A fourth method, also interpretative, uses the information contained upon monochrome air photographs. Without going into the details of air-photo interpretation, an appreciation of relief can be obtained from stereoscopic examination. The tone of the print together with its texture, gives a good impression of the variation of the colour of the soil or of the vegetation growing upon it. Consequently, a skilled interpreter can see the distribution of both vegetation and soil and mark the boundaries on the photograph. This is made possible by the reaction of plants to different soil conditions. It is then an essential second stage of the survey to go out into the field and to examine the soil within the areas determined from the air photograph to see if they are truly representative of the vegetation patterns which they support. Recent developments in air photography have used colour and other sensory methods to map soil distributions.

A skilled soil surveyor uses all the information which is available. This includes not only the soil itself, but the breaks and changes in the slope of the land, the geology of the parent material, the vegetation which grows naturally, including hedgerows in arable areas, as well as the present land use. However, in the latter case utilisation can be dictated by purely economic considerations which are not always in the best interests of soil conservation and good farming.

Once the different soil-mapping units have been mapped it is usual for the surveyor to dig a hole, make a description and sample the horizons of the proposed profile at selected sites. The samples are placed in waxed paper or polythene bags, labelled and brought back to the laboratory, where analyses can be made of the samples to support the decisions about classifications made in the field. The profile description is recorded on a field soil description sheet with observations conveniently listed under the headings of site details and profile details:

A Site details
  Profile number
  Locality
  Map reference
  General group
  Series and type
  Elevation, relief and aspect
  Drainage of site and profile
  Parent material
  Natural vegetation or agricultural practice
  Climatic and weather conditions

The features listed are clearly designed to give information about the area in which the soil profile is situated, and the different sections should be answered as fully as possible by the surveyor at the time of description.

B Profile details
  Horizon depth and clarity
  Colour
  Texture
  Stoniness, size and amount
  Structure
  Consistence
  Organic matter
  Roots
  Moisture conditions
  Fauna
  Concretions of pedological origin
  Presence of carbonates
  pH

The above details are given for each of the soil horizons found to occur within the soil profile. The full range of terminology is given, in the case of Britain, by the field handbook of the Soil Survey of Great Britain. The terminology used by other countries is similar, but care must be exercised to check the exact use of the terms employed. Measurement in centimetres for the depth of soil horizons is now commonplace and the use of the Munsell colour charts is widely accepted for the description of soil colours. An international range of textures has been agreed upon, and is incorporated into most soil survey systems, but some other measurements for particle size are also used (p. 8).

Once the various horizons of a profile have been described and identified, they are referred to by a system of nomenclature using capital

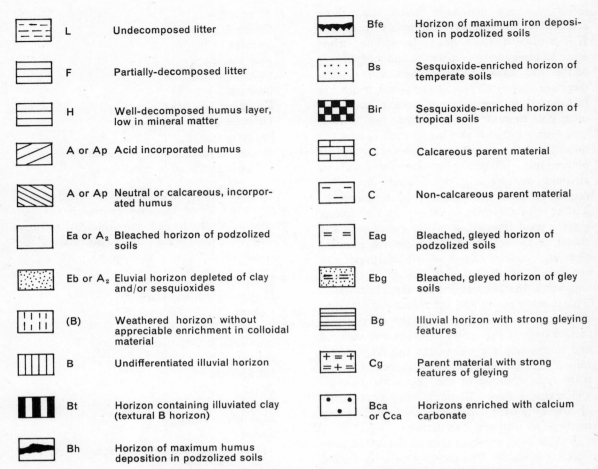

	L	Undecomposed litter
	F	Partially-decomposed litter
	H	Well-decomposed humus layer, low in mineral matter
	A or Ap	Acid incorporated humus
	A or Ap	Neutral or calcareous, incorporated humus
	Ea or A₂	Bleached horizon of podzolized soils
	Eb or A₂	Eluvial horizon depleted of clay and/or sesquioxides
	(B)	Weathered horizon without appreciable enrichment in colloidal material
	B	Undifferentiated illuvial horizon
	Bt	Horizon containing illuviated clay (textural B horizon)
	Bh	Horizon of maximum humus deposition in podzolized soils
	Bfe	Horizon of maximum iron deposition in podzolized soils
	Bs	Sesquioxide-enriched horizon of temperate soils
	Bir	Sesquioxide-enriched horizon of tropical soils
	C	Calcareous parent material
	C	Non-calcareous parent material
	Eag	Bleached, gleyed horizon of podzolized soils
	Ebg	Bleached, gleyed horizon of gley soils
	Bg	Illuvial horizon with strong gleying features
	Cg	Parent material with strong features of gleying
	Bca or Cca	Horizons enriched with calcium carbonate

5.6 Symbols used to represent soil horizons throughout this book. These symbols are not used by all soil scientists in their interpretation of soil profiles. Consequently, not all the examples used in this book have them. Combinations of symbols indicate a horizon with features common to both

letters with subscripts. This system was originated by Dokuchaiev who simply labelled his soil horizons A, B, C, but eventually these symbols came to have a genetic significance as can be seen from the explanation of the symbols in Fig. 5.6. The application of this notation can be seen in various profiles described later in this book.

The tools used by soil surveyors are mostly fairly straightforward and can be supplied by most competent tool shops, or laboratory suppliers (Fig. 5.7). For rapid inspection of the soil, a screw auger is probably the best tool in that it can be used to remove successive samples from a small hole. Although these small samples are useful for recognition of colour and texture, no appreciation of structure can be obtained, and

in dry weather it can be difficult to extract samples of sandy soils from the hole. Similar comments can be made about the Jarrett auger, a larger tool with a 10 cm. diameter bucket. It gives a bigger sample than the screw auger but structure is again difficult to assess as the sample is disrupted as it is brought up. A power-auger can be fitted to a vehicle such as a Landrover, but these are expensive and they cannot be used successfully in very stony soils. The larger size of the cores of the power-auger does enable an assessment of soil structure to be made on the sample when the cylindrical core which is removed is split down the middle.

Normally, the digging of a profile pit necessitates the use of a strong spade, and in some cases

5.7 Tools of the soil surveyor

a pick-axe is essential. A pit should be large enough to obtain the necessary information, and yet not so large that effort is wasted in the removal of a great weight of soil. Most soil pits are 1 m. wide and 1.5 m. long at the surface, with steps down to successively deeper levels. The face to be described is usually chosen so that it is well lit to facilitate description and photography. A trowel, either an ordinary gardener's trowel, or a special sampling trowel, is useful for removing representative samples from the face of a profile pit. The trowel is useful, too, for 'facing-up' the pit side before the process of description and sampling begins. This is the process of removing the smeared soil and exposing fresh structures and surfaces.

Other necessary aids to soil profile description are a clearly marked steel measuring tape, a small kit to determine the pH value of the soil at each horizon, and a small bottle of 10 per cent hydrochloric acid with which to test for the presence of carbonates. A form of map case which can be used for writing upon in the field is essential as this can hold maps and field soil description sheets. Many surveyors find a pocket notebook is useful for recording details of sections, vegetation, land use and other relevant information not easily recorded on the field map.

6 SOILS OF THE HIGH LATITUDES

The largest areas of *tundra soils* are found in the northern hemisphere surrounding the Arctic Ocean. The distribution of these soils in Eurasia extends northwards of a line from the extreme north of Norway, across the north of Siberia approximately along the line of the Arctic Circle, reaching the coast again in the region of the Kamchatka Peninsula (Fig. 6.1). A similar distribution is seen in North America where the line extends along the arctic coast of Alaska to the Great Bear Lake, the southwestern shore of Hudson's Bay, crossing the Labrador–Ungava Peninsula to reach the Atlantic coast. Altogether approximately 4 per cent of the land area of the world has tundra soils.

The tundra is characterised by a severe climate with long, bitterly cold winters and short cool summers. Tundra climate (ET) as defined by Köppen has two to four months only of average temperatures above freezing, and a killing frost

6.1 Distribution of permafrost in the tundra regions of the northern hemisphere

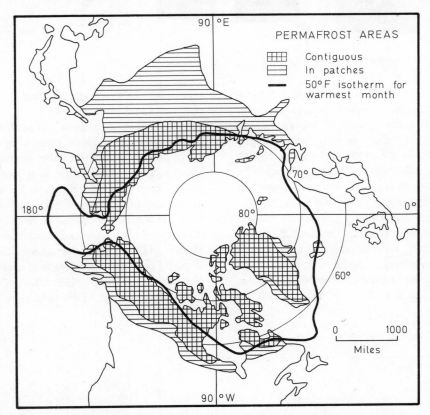

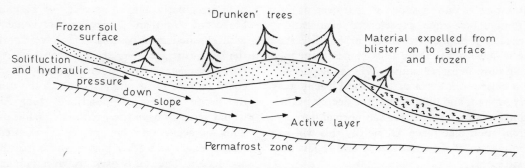

Frozen soil surface

'Drunken' trees

Material expelled from blister on to surface and frozen

Solifluction and hydraulic pressure

down slope

Active layer

Permafrost zone

6.2 A soil blister or pingo; water from the slope descends under hydraulic pressure and lifts the frozen surface soil in a blister up to 7 metres high

can occur at any time. Winter temperatures are as low as −35 °C or − 40 °C. Tundra regions also have a low precipitation, usually 250 to 300 mm. per annum.

In the southern hemisphere tundra soils are less extensive, but they do occur on areas of Antarctica not covered by ice, the elevated areas of the Andes, the Great Dividing Range in the southeast of Australia, and in the Southern Alps of New Zealand. Tundra soils can be seen at high altitude on mountains in other parts of the world, but their extent is strictly limited to the higher peaks of mountains.

The parent materials of soils in the tundra are diverse, ranging in age from Pre-Cambrian shield areas of Canada and Siberia to the Cretaceous of Spitzbergen and the boulder clay and alluvium of more recent time. The local rocks will be slowly weathered by physical processes chiefly, and weathering will tend to be concentrated in the periods when an alternation of freezing and thawing takes place.

Below the soil in the tundra regions is a layer of permanently frozen ground, known as the *permafrost*. It represents the soil and rock which is not thawed out during the brief period of the arctic or antarctic summer. The soil which does become thawed during the summer is saturated with water which cannot escape downwards by percolation because of the permafrost layer beneath. In these conditions the soil becomes so wet that it can flow gently downslope, a process which is called *solifluction*. As a result of pressures generated when the surface horizons freeze, the still liquid material lying above the permafrost sometimes erupts on the surface forming mounds

known as *pingos* (Fig. 6.2). Thus there is a continuing process of redistribution of soil matter associated with each period of freeze and thaw. The general result is for the material to move downslope and at the same time to become roughly sorted with the finer material occupying the lower parts of the landscape leaving thin soils with coarse rock fragments on the hills.

The 'cold deserts' are those parts of the world in polar regions and amongst high mountains which are not currently covered by glacier ice. They are treeless with the tree line marking their boundaries. In the tundra, the depth of rooting for trees is limited by the permafrost, and although water is plentiful during periods of thaw, for much of the year it is unavailable because it is frozen. Growing vegetation can therefore experience a physiological drought, which is estimated as a deficiency of up to 170 mm. at Barrow, Alaska, an environment which can be classed as semi-arid. At the boundary of the tundra where sheltered sites permit, spruce, larch, pine, birch, aspen, willow and mountain ash will grow. However their growth is slow, and they are frequently small and deformed because of the severe climate. On the tundra itself, heather, arctic blueberry, as well as flowering plants such as buttercups and mountain avens, lichens and mosses are the most common where other plants cannot survive.

Although the rate of plant growth is slow, so is the rate of decomposition of organic matter, consequently there is a slow accumulation of organic matter in these soils. Soil formation is essentially restricted to the summer when the upper layers of the soil thaw. Precipitation is

41

slight, either as rain or snow, so that even at the low temperature then experienced evaporation can exceed precipitation. In all probability the water which leaches these soils comes from the melting snow in early summer. Where the permafrost is close to the surface, downward leaching is clearly limited, and only where it is deeper in the soil and the soil is permeable can normal horizon differentiation take place. Obviously this will be on the margin of the tundra zone and in favourable sites only.

Any tendency towards horizon development is countered by the process of churning which occurs in the 'active' layer. Because the soil freezes from the surface downwards, a saturated layer of soil is trapped between it and the permafrost. As there is expansion in the process of freezing, pressures are built up which eventually rupture the surface, spilling the lower material on to it. This churning process is referred to as *cryoturbation*. This process with the solifluction mentioned previously, inhibits the development of horizons which are the characteristic features of the soils of temperate climates.

The sorting processes produce a soil catena which has been described from Spitsbergen by Smith (Fig. 6.3). Upper slopes are characterised by frost shattering, and screes at angles of 34° develop below the outcrop of solid rocks. The movement downslope of loose materials merges into a striped pattern of gravel with finer soils between (Plate 1). The stripes occur on slopes up to 25°. Lower-angled slopes of 12°–4° are characterised by soils showing signs of solifluction

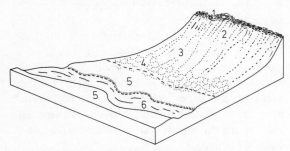

6.3 A block diagram illustrating relationships of soils found in tundra regions. 1. Frost-shattered rock on ridge crests. 2. Scree on steep slopes over 25°. 3. Stone stripes on slopes 12°–24°. 4. Arctic gley soils on fine solifluction material (polygon soils). 5. Arctic brown soils on river terrace materials. 6. Alluvial material in valley bottoms

with irregularly shaped terraces of bare mud. Slopes of less than 4° develop polygonal soils with elevated mud centres. The lowest parts of the landscape are covered with peat deposits.

If a case is to be made for a mature soil which could be said to be characteristic of the tundra, the arctic brown soil described from Alaska would seem to have the necessary qualifications. This soil occurs on ridge tops, escarpment edges, terrace edges and other places where free drainage occurs. The soil is dark brown in colour and has fragile fine-crumb structure in the A horizon. A dark yellow-brown B horizon of sandy loam overlies a horizon of shattered sandstone bedrock.

Profile of an arctic brown soil from Alaska
(Parent material – sandstone)

2–0 cm.	Black organic layer
0–17 cm.	Dark brown (7.5YR3/2) sandy loam with single grain to fine crumb structure. Very loose and friable
17–36 cm.	Dark yellow-brown (10YR3/4) sandy loam with single grain to fine crumb structure. Loose and friable
36–60 cm.	Dark yellow-brown (10YR4/4) sandy loam. Firm but friable
60–80 cm.	Very dark grey-brown (2.5Y3/2). Loamy sand with small weakly indurated aggregates. Numerous small rock fragments
80 cm.+	Shattered bedrock with numerous sandstone fragments

(After Tedrow)

Note. Symbols following colour descriptions refer to the notation used in the Munsell Soil Colour Charts (p. 37).

The area of freely drained soils is limited. More typical of the tundra is a gley or peaty gley soil formed with conditions of poor drainage. On clay-rich parent materials, a Russian author has described an arctic gley soil from Wrangel Island where above the permafrost a gley horizon of 44 cm. underlies this organic-rich (5 per cent) A horizon.

Profile of an arctic gley soil from Wrangel Island
(Parent material – river terrace deposit)

2–0 cm.	Peaty, part-decomposed plant remains
0–1 cm.	Brown clay loam with fine (crumb?) structure and roots
1–44 cm.	Grey light-blue with yellow horizontal streaks of clay loam with fine structures, porous, some plant roots extending to 38 cm.
44 cm.+	Permafrost

(After Svatkov)

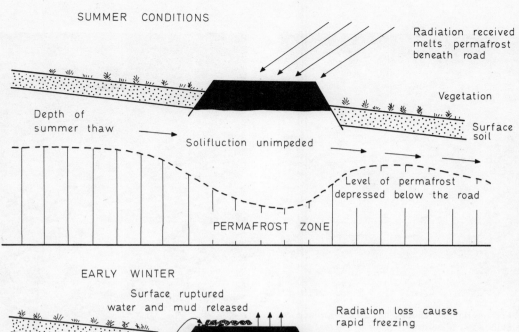

SUMMER CONDITIONS

Radiation received
melts permafrost
beneath road

Depth of
summer thaw

Solifluction unimpeded

Vegetation

Surface
soil

Level of permafrost
depressed below the road

PERMAFROST ZONE

EARLY WINTER

Surface ruptured
water and mud released

Radiation loss causes
rapid freezing

Hydraulic pressure
down slope

Rapid freezing below
road blocks solifluction flow
down slope

Line of summer
melting

PERMAFROST ZONE

6.4 The effect of a road upon soil in the tundra

The relationships of the arctic gley soil and the arctic brown soils are shown in Fig. 6.3. Other soils which have been described from arctic tundra regions are transitional to the podzols, gley podzols and peat. Salt-enriched solonchak soils also occur where salts accumulate at the foot of slopes or where salt spray is driven inland and accumulates in the soils. Thus in spite of the severity of the climate and limitation on soil-forming processes there is a considerable range of soils to be seen in tundra regions. Tundra soils are variously allocated in the American Classification to cryic great groups of the orders such as the Entisols and Spodosols, e.g. Cryaquent – polygonal soils; Cryaquod – ground-water podzol.

Interest in tundra soils has been stimulated by strategic needs during the last three decades in both Russia and North America. Also the extraction of valuable minerals from the arctic regions necessitated the building and maintenance of road and rail communications. The engineering problems encountered led to considerable research which has revealed additional information about the area and its soils (Fig. 6.4).

The importance of the study of tundra soils lies in the fact that during the Pleistocene their extent was much greater and included much of North America, Europe and the British Isles where relic features of tundra soils are widespread. Features variously known as involutions, convolutions or festoons which can be attributed to the churning movements of the former active layer can be seen in sections revealed in quarries (Fig. 6.5). The upper layer of the permafrost

6.5 The effect of a periglacial climate can be seen in features known as involutions or festoons

6.6 An ice wedge pseudomorph. This feature once occupied by ice has been filled by different material A which distinguishes it from the surrounding chalk B

became compacted by ice growth and infilling by silt to form an undurated horizon which has been described from many soils in Scotland and Wales. Wedges of ice which penetrated deeply into the weathered mantle during the last cold period eventually melted out and the space left became

6.7 Frost polygons. Constant freeze-thaw activity results in sorting coarse from fine material. The coarse material is pushed into a polygonal pattern

infilled with material of a different nature. These infillings, fossil ice wedges, can also be seen in quarry sections of gravel, sand or chalky material (Fig. 6.6). An example of polygonal ground formed during a previous cold period is exposed on the shores of Cardigan Bay, where it was covered by a later deposit which the sea is eroding once again (Fig. 6.7).

An understanding of tundra soils is necessary to interpret the features seen in the sub-soils of many British and European soils, the distribution of whose parent materials is impossible to explain without reference to periglacial climate. As the surface horizons may have been changed by the current soil-forming factors so the soils are appropriate for the area, but below is concealed the evidence of an earlier period of different climate and soil-forming processes.

1 Stone stripes, Iceland. Regular freeze-thaw cycles have separated material of different sizes into parallel lines, a feature found particularly on sloping ground. **2 Humus-iron podzol,** Norfolk, U.K. This profile is developed in acid, freely drained sands and gravels beneath a 'heath' type of vegetation. Eluvial and illuvial horizons are clearly seen. **3 Humus podzol,** Drenthe, Holland. This profile is developed in dune sands lacking in iron, consequently only a Bh horizon develops. Multiple profiles have developed where further sand has accumulated. **4 Gley podzol,** Lincolnshire, U.K. Poorly drained conditions cause the mottled appearance with areas of weakly cemented orange-brown sand. The surface soil is rich in organic matter and the profile has developed in blown sand.

5 Peaty gleyed podzol, Breconshire, U.K. Higher rainfall and peaty tendency cause poor drainage in surface horizons of soil. A thin, often convoluted, iron pan develops below which the soil is freely drained. **6 Brown podzolic soil,** Glamorganshire, U.K. Enriched by humus in the surface horizon, these soils lack an eluvial Ea horizon. The B horizon is highly coloured and the soil is developed on sloping sites on acid parent materials. **7 Acid brown soil,** Derbyshire, U.K. This profile is developed upon base-deficient parent materials and does not show any evidence of accumulation of clay or of iron or humus in the (B) horizon. **8 Leached brown soil,** Derbyshire, U.K. Developed from fine-grained, slightly calcareous parent material, these soils are first leached, then movement of clay into the Bt horizon begins.

9 Brown earth with gleying, Nottinghamshire, U.K. Super-imposed upon the features of a leached brown soil are the pale colours indicative of slow drainage. **10 Grey forest soil,** U.S.S.R. The development of a pronounced illuvial horizon by clay and humus movement has left the A horizon structures with a coating of fine silica grains which give the grey colour. **11 Red Mediterranean soil,** Toledo Province, Spain. Dev-eloped from Silurian shales, this profile has the pronounced reddened, clay-enriched Bt horizon typical of these soils. **12 Red-brown earth,** Barossa Valley, South Australia. These soils are characterised by a red-brown surface soil below which is a darker, clay-enriched Bt horizon the lower parts of which are calcareous.

13 Calcareous concretion, South Australia. The lower horizon has become impregnated with a rock-like accumulation of calcium carbonate (the knife rests upon it). Erosion of the overlying soil exposes the 'crust' of calcium carbonate at the surface. **14 Yellow podzolic soil,** N.S.W., Australia. Strongly weathered and leached profile develops in low-lying situations in the landscape where moister conditions prevail. **15 Chernozem,** U.S.S.R. The deep humus-rich A horizon of this 'typical' chernozem is about 1 m. in depth. Krotovinas can be seen in the lighter-coloured C horizon. The horizon of accumulation of calcium carbonate Cca, deeper in the pit, cannot be seen here. **16 Southern chernozem,** U.S.S.R. The A horizon is not so deep nor so rich in humus as in the 'typical' chernozem. Note the calcium carbonate accumulation in the C horizon.

17 **Chestnut soil,** U.S.S.R. Drier conditions and less organic matter give these shallow soils a lighter colour. The calcium carbonate accumulation is closer to the surface as leaching is less effective. **18 Desert soil,** N.S.W., Australia. Formed from mineral material and lacking in organic matter, this soil is little more than an accumulation of alluvial material washed from adjacent higher ground. **19 Ferrallitic soil,** Misuku, E. Africa. Freely drained sites in the humid tropical regions develop deep red soils which are acid in reaction. Silica is removed from the clay minerals. Quartz is relatively stable. Iron and aluminium sesquioxides may form a substantial part of the mineral matter. **20 Yellow latosol,** Queensland, Australia. In moister situations yellow iron compounds are formed. Other characteristics are similar to those of red ferrallitic soils.

21 Vertisol, N.S.W., Australia. The lowest parts of the landscape in tropical and sub-tropical areas have dark coloured, base-rich soils with montmorillonitic clays. **22 Gilgai,** Narrabri, N.S.W., Australia. The physical properties of the clay in these soils enables them to crack widely. Material falls down the cracks, and the soil gradually inverts itself. It also is forced into humps and hollows with an amplitude of 1 to 2 metres.

23 Krasnozem, Childers Plateau, Queensland, Australia. Deep, friable, red, loamy soils developed from base-rich parent materials occur in the sub-tropics. As the high iron content restricts eluvial movement, horizon development is not obvious. **24 Pisolitic laterite,** Victoria Point, Queensland, Australia. Lateritic concretions can take the form of rounded nodules consisting mainly of iron oxides.

25

26

27

28

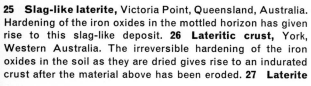

25 Slag-like laterite, Victoria Point, Queensland, Australia. Hardening of the iron oxides in the mottled horizon has given rise to this slag-like deposit. **26 Lateritic crust,** York, Western Australia. The irreversible hardening of the iron oxides in the soil as they are dried gives rise to an indurated crust after the material above has been eroded. **27 Laterite** **exposure,** York, Western Australia. Erosion has left laterite cappings to the interfluves in the York area of Western Australia. The ironstone crust and pallid horizon can be seen clearly. **28 Rendzina,** Transylvania Alps, Romania. Shallow, humus-rich soils over calcareous parent materials are referred to as rendzinas.

29

30

31

32

29 Gley soil, Derbyshire, U.K. Poorly drained conditions in the soil cause chemical reduction of iron compounds; mottling with grey and orange colours are typical. **30 Solonchak,** Buzau, Romania. An alluvial soil which has become salinized. This detailed photograph shows pores and channels in the soil filled with white salts. Efflorescences of salts also occur on the surface of the soil. **31 Solonetz,** U.S.S.R. Partly leached of salts, this soil profile demonstrates the rounded tops of the columnar structures associated with solonetz. The topsoil shows a distinct lack of structure. **32 Peat,** Breconshire, U.K. On upland, high rainfall areas in western Britain, accumulation of peat has been encouraged to form the 'blanket bog'. These organic soils have developed upon acid shales with strong gleying conditions in the mineral material beneath.

7 SOILS OF THE MID-LATITUDES, COOL CLIMATES

Four zonal soils can be considered as typical of the cool temperate climates: podzols, brown earths, leached soils and grey soils. Although there is considerable climatic variation in the mid-latitude areas of cool climate, there is sufficient precipitation to maintain an overall downward leaching of any soluble soil constituents. The podzol is typical of more northern areas where it is associated with the boreal forest; however, podzols can also be seen where heath plants occupy infertile sandy or gravelly areas within the deciduous forest areas. Brown earths and grey-brown podzolic soils are normally associated with the deciduous forest, and the grey soils with the forest-steppe transition.

Within this broad zone of soil formation, there are many divergences from the zonal types. These divergences occur because of the interplay of parent material, vegetation, accumulation of organic matter, soil drainage or stage of maturity. Thus within this zone of intrazonal rendzinas, rankers, gley and organic soils can be identified in addition to the azonal skeletal soils and those formed from recent alluvium. These soils have in common the fact that they are all leached to some extent. The different zonal soil groups are differentiated by the visual appearance of their profiles, though sub-groups may need some laboratory investigation for their accurate identification.

Podzols

These soils are characterised by the presence, just below the surface, of an ashy-coloured horizon from which they derive their Russian peasant name. On world soil maps podzols are shown as extending in a circumpolar belt approximately from the Arctic Circle southwards to 50 °N in Eurasia, and slightly further south, to the latitude of the Great Lakes in North America (Fig. 5.1).

South of this main circumpolar zone, podzols are restricted to areas where a combination of parent material, climate and vegetation is favourable for their development. Podzols are found also in limited areas of Australia and South America, and where higher mountains or suitable parent materials occur. (A form of podzol can also be seen in the tropics where deep profiles have developed on alluvial sands.) Within the main podzol zone in the northern hemisphere, large areas of the earth's surface are mantled with Pleistocene fluvio-glacial deposits. Although podzolization proceeds most rapidly upon permeable sands and gravels, soils of this type can be found on a wide range of different parent materials including sandstones, siltstones, clays, and in certain circumstances even soils partly derived from limestones.

The climates associated with the development of podzol soils have cold winters and short, warm summers. In European Russia, these soils are frozen for at least five months of the year, and during the brief summer temperatures only reach between 15 °C and 19 °C. In this typically continental climate, winter precipitation is in the form of snow, and about half the annual precipitation is in the form of summer rain. The total amount of precipitation is not great, 500–550 mm. are stated as average for Eurasia, but amounts of up to 1000 mm. occur in North America. In terms of Thornthwaite's evapotranspiration figures, the main podzol zone lies between the 400–500 mm. lines, and according to Köppen the southern boundary is approximately that of the 'cold zone' which experiences a temperature of over 10 °C in the warmest month and below −33 °C in the coldest month. Podzols are found predominantly in Köppen's humid microthermal climates (Dfa, Dfb, Dwa, Dwb) as well as in sub-arctic climates (Dfc, Dwc, Dwd) and marine west coast climates (Cfb). The freezing conditions of winter months in the colder areas largely inhibit the soil-forming tendencies, but with the snow-melt of spring, considerable moisture is available, and leaching and gleying of the soil profile occurs.

Podzols (Orthods) are soils associated with heath and the coniferous forests of northern latitudes, alternatively known as the boreal forest or taiga. In Europe this forest is composed of spruce, pine, larch and birch; in Siberia, the Dahurian larch is dominant in the areas of the most extreme climatic ranges; and in the Far East, fir becomes co-dominant with spruce. The number of species in the American boreal forests is greater, with spruce and fir growing on the better drained soils and black spruce and tamarack on the poorly drained soils. Burnt areas are colonised with jack pine, birch and aspen, while western forests are characterised by lodgepole pine and alpine fir. As the ecotone with the tundra is approached, the boreal forest thins and a discontinuous ground flora of crowberry and bilberry occurs as well as lichens such as *Cladonia*. South of the main podzol zone where podzols may form beneath a mor humus derived from heath plants which have colonised acid sandy soils, this organic matter is noticeably deficient in bases and plant nutrients which are being returned to the soil. The acidity and the lack of light penetration to the forest floor reduces the number and range of organisms which should cause breakdown of plant material and humification. Breakdown of plant material is slow, mainly being achieved by the activity of fungi, with fauna such as mites being of secondary importance. The humus form is that of mor, in which the litter, fermentation and humus horizons can usually be identified. As earthworms are absent, incorporation into the soil is slow and thus a mat of organic material is gradually accumulated on the surface of the mineral soil.

A podzol soil is one in which a redistribution of the soil constituents has taken place by the downward-percolating rainwater. This water contains decomposition products of the organic matter described above, and as it passes through the soil it can dissolve or extract any further soluble constituents. It has been shown that in acid conditions, such as are found in a podzol,

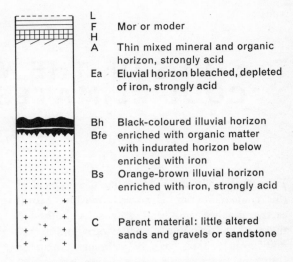

L F H	Mor or moder
A	Thin mixed mineral and organic horizon, strongly acid
Ea	Eluvial horizon bleached, depleted of iron, strongly acid
Bh Bfe	Black-coloured illuvial horizon enriched with organic matter with indurated horizon below enriched with iron
Bs	Orange-brown illuvial horizon enriched with iron, strongly acid
C	Parent material: little altered sands and gravels or sandstone

7.1 Diagrammatic profile of a humus-iron podzol

fine particles of organic matter are capable of forming organo-mineral complexes which are carried down the profile. Several different explanations of the way in which iron can be moved from the A and Ea horizons have been put forward including chemical, physical and biological processes. However, although much work has been done on this problem, there is as yet no unanimous agreement about what is involved. It is certain that constituents are removed and redeposited lower in the soil profile. Because of the removal of constituents, notably iron and aluminium, and the breakdown of clays, there is a relative increase in the amount of silica remaining in the Ea horizon and an increase in the amount of sesquioxides of iron and aluminium in the Bs horizon. This process results in the distinctive horizons of the podzol profile (Fig. 7.1). However, there are a number of variations which may be seen in the field examination of these soils. On lowland sites it has been suggested that some of these variations may be linked in a maturity sequence. This begins with a podzolized acid brown soil with weakly developed Ea and Bs horizons, continues with an iron podzol in which the oxides of iron have been moved to form the Bs horizon, and culminates in the most mature soil, the humus-iron podzol (Fig. 7.2 and Plate 2). In this soil an accumulation of organic material, a Bh horizon, occurs in and above the Bfe and Bs horizons, suggesting that

46

the organic matter is arrested in its progress down the profile by the increasing accumulation of iron. This stage is associated with a heathland vegetation in northwestern Europe. Parent materials lacking in iron cannot develop a humus-iron podzol profile, and certain dune sand deposits in Holland have well formed humus podzols (Humods) in which the Bfe and Bs horizons are absent (Plate 3). Horizon sequences are given in Table 7.1.

Profile of a humus-iron podzol, Derbyshire, England (Parent material – Bunter Pebble Beds)

L		Discontinuous litter of beech, oak and larch leaves
F	4–2.5 cm.	Comminuted, darkened and partly decomposed leaves
H	2.5–0 cm.	Black amorphous organic matter with a scatter of bleached sand grains
A	0–13 cm.	Very dark grey (10YR3/1) structureless sand with bleached grains; slightly stony with moderately high organic matter content; merging boundary
Ea	13–30 cm.	Grey (10YR5/1) structureless, slightly stony sand with bleached grains and low organic matter content; abrupt boundary
Bh	30–31 cm.	Black (10YR2/1) indurated sand and stones with illuvial humus and iron accumulation; abrupt boundary
Bfe	31–32 cm.	Dark brown (7.5YR3/2) indurated sand and stones with illuvial iron accumulation; narrow boundary
Bs	32–40 cm.	Reddish-yellow to strong brown (7.5YR6/6 to 5/6) very stony compact sand with no apparent organic matter
C	40 cm.+	Bunter Pebble Beds

(After Bridges)

Where a high ground-water occurs in a pervious parent material undergoing podzolization, the normal processes are modified. A ground-water podzol (Aquod) forms which can have B horizons enriched with iron or organic matter or both. The oxidation and reduction caused by alternate aerobic and anaerobic conditions gives the B horizon a mottled appearance with patches of weakly cemented orange-brown sand (Fig. 7.3 and Plate 4). In the podzol zone where wet conditions coincide with impervious parent materials gley or peaty gley soils usually result on lowland sites (Fig. 7.4). It should be noted that the Russian podzol includes many soils which the western pedologists would classify as hydromorphic soils, hence the great

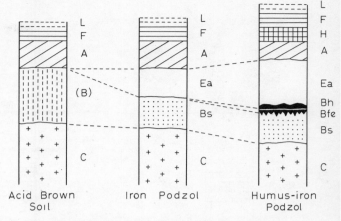

7.2 Maturity sequence of soils from acid brown soil to humus-iron podzol (after Mackney)

Table 7.1 *Sub-groups and horizon sequence of podzol soils*

Sub-groups	Horizon sequence
Lowland:	
Iron-podzol	L F H A Ea Bs C
Humus-podzol	L F H A Ea Bh C
Humus-iron podzol	L F H A Ea Bh Bfe Bs C
Ground-water podzol	L F H A Ea Bh/Bsg Cg
Upland:	
Brown-podzolic	L F H A Ea (B) C
Peaty gleyed podzol	Peat Ag Eag Bfe Bs C
Gley podzol	Peat Ag Eag Bsg Cg

7.3 Diagrammatic profile of a ground-water podzol

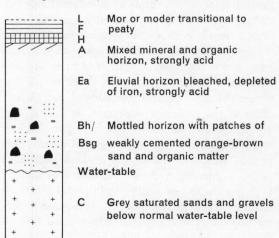

L F H A	Mor or moder transitional to peaty
	Mixed mineral and organic horizon, strongly acid
Ea	Eluvial horizon bleached, depleted of iron, strongly acid
Bh/ Bsg	Mottled horizon with patches of weakly cemented orange-brown sand and organic matter
Water-table	
C	Grey saturated sands and gravels below normal water-table level

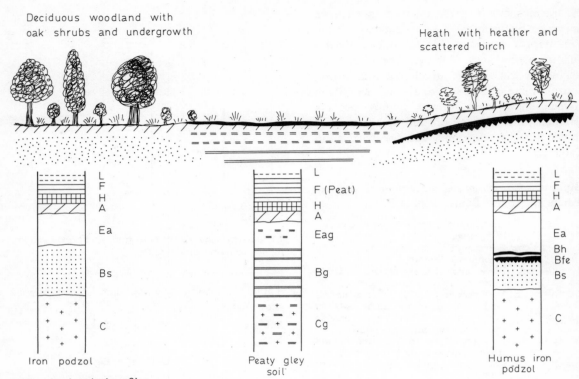

Deciduous woodland with oak shrubs and undergrowth

Heath with heather and scattered birch

L F H A		L		L F H A
Ea		F (Peat)		Ea Bh Bfe
Bs		H A		Bs
C		Eag		C
		Bg		
		Cg		

Iron podzol

Peaty gley soil

Humus iron podzol

7.4 Lowland podzol profiles

extent of podzol soils in northern Russia. In western Europe the podzol is thought of as a freely drained soil which occurs predominantly on lowland heaths.

Strong leaching and podzolization also occur on the uplands of western Europe where a form of podzol can be seen to have developed in a high rainfall regime. In an environment which has greater cloudiness, and a decreased rate of evapotranspiration, accumulation of an acid organic mat is encouraged. Gleying results below the mat and an Eag horizon develops. A thin iron pan, which resembles a walnut shell in appearance occurs below the Eag horizon. Water accumulating upon this pan adds to the gleying effect of the saturated superficial horizons (Fig. 7.5 and Plate 5). Below the iron pan, and in striking contrast, the soil is freely drained. This profile, which resembles the lowland podzol, is called a peaty gleyed podzol, and it is usually associated with peaty gleys and peats. In western Britain peaty gleyed podzols (Placaquod) frequently occur on the imperfectly drained sites on the upper convexity of hills (Fig. 7.6).

There have been difficulties in delimiting and classifying the soils which have minimal development of podzol features. Often the evidence of podzolization is shallow and easily destroyed by agriculture. As there is no real break between the

7.5 Diagrammatic profile of a peaty gleyed podzol

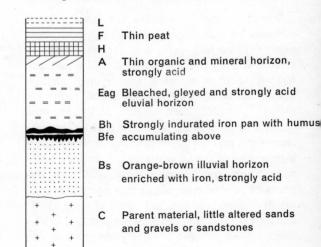

L
F   Thin peat
H
A   Thin organic and mineral horizon, strongly acid

Eag Bleached, gleyed and strongly acid eluvial horizon

Bh   Strongly indurated iron pan with humus
Bfe accumulating above

Bs   Orange-brown illuvial horizon enriched with iron, strongly acid

C   Parent material, little altered sands and gravels or sandstones

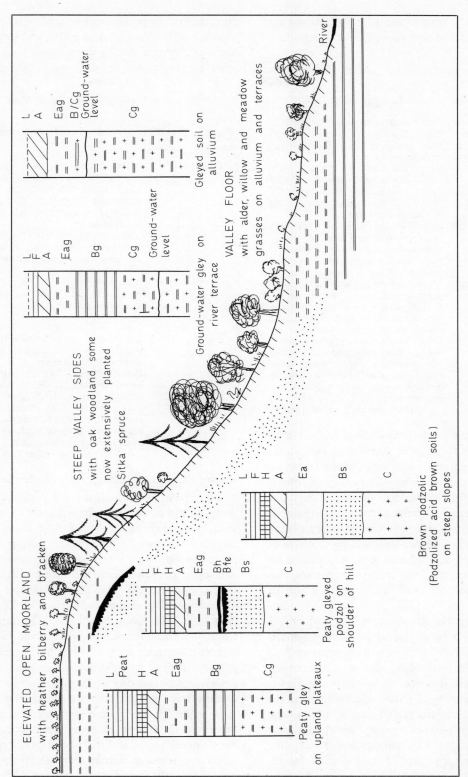

ELEVATED OPEN MOORLAND
with heather bilberry and bracken

STEEP VALLEY SIDES
with oak woodland some
now extensively planted
Sitka spruce

VALLEY FLOOR
with alder, willow and meadow
grasses on alluvium and terraces

River

Ground-water gley on
river terrace

Gleyed soil on
alluvium

L
F
A
Eag
Bg
Cg
Ground-water
level

L
A
Eag
B/Cg
Ground-water
level
Cg

Brown podzolic
(Podzolized acid brown soils)
on steep slopes

L
F
H
A
Ea
Bs
C

Peaty gleyed
podzol on
shoulder of hill

L
F
H
A
Eag
Bh
Bfe
Bs
C

Peaty gley
on upland plateaux

L
Peat
H
A
Eag
Bg
Cg

7.6 Upland podzol profiles

podzols and the brown earths, some overlapping of the characters of both groups must be expected.

Profile of a peaty gleyed podzol, Shropshire, England
(Parent material – hard Longmyndian shales)

F	10–7·5 cm.	Partly decomposed plant remains
H	7·5–0 cm.	Black amorphous organic matter with pieces of charcoal and many heather roots
Eag	0–13 cm.	Very dark grey-brown sandy loam with coarse sub-angular blocky peds; numerous well-weathered stones; few fissures or pores; abundant fine roots
Bh/Bfe	13–14 cm.	Soft 0·25 cm. layer of very dark brown organic-rich material overlies a strongly cemented 0·25 cm. iron pan; marked concentration of fine roots at the top of the pan
Bs	14–28 cm.	Strong brown gravelly loam; numerous stones up to 8 cm. diam.; friable; weakly cemented, breaking to crumb and fine sub-angular blocky peds; frequent fine roots
C	28 cm.+	Dark grey-brown gravelly sandy loam; weakly cemented; no fissures; but abundant pores; roots common; some dead; below 45 cm. the horizon merges into a loose, extremely stony layer

(After Mackney and Burnham)

Brown podzolic soils are identified in the north-eastern area of the U.S.A. These soils lack an Ea horizon, but possess a B horizon which shows no increase in content of clay-sized material. The B horizon is highly coloured in the upper part and this colour gradually fades with increasing depth (Plate 6). Whilst these soils are obviously closely related to the acid brown soils and particularly to the podzolized acid brown soils, a case can be made for retaining both acid brown soils and the brown podzolic soils as separate items in classifications.

Soils of similar nature have been identified over extensive areas of sloping land in the western parts of Britain and Europe where names such as podzolized sol brun acide and sol brun ocreux have been used. In these regions these slope soils, developed upon hard Palaeozoic sediments, are often the only freely drained soils in a landscape dominated by peaty gley soils or upland gley podzols (Fig. 7.6). The latter group of soils, like the ground-water podzol of the lowlands, has its lower horizons affected by the presence of water.

In addition, the presence of an acid peaty humus together with bleached sand grains and gleying in the Eag horizon distinguishes these soils from other poorly drained upland soils (Table 7.1).

Profile of a brown podzolic soil, New York State, U.S.A.
(Parent material – glacial till and outwash material)

L, F, and H		Matted mor humus
A	0–2·5 cm.	Greyish-brown (10YR4/2) very fine sandy loam with weak fine crumb structure; very friable; white flecks of incipient eluvial horizon Ea may occur
Bs$_1$	2·5–20 cm.	Yellowish-brown (10YR5/6) very fine sandy loam with weak fine crumb structure; very friable; some stone fragments
Bs$_2$	20–45 cm.	Yellowish-brown (10YR5/4) very fine sandy loam with weak medium crumb structure; friable; some stone fragments
Bs$_3$	45–65 cm.	Light yellowish-brown (10YR6/4) fine sandy loam or loam with weak coarse crumb structure; friable; some stone fragments
C	65–200 cm.	Pale brown (10YR6/3) to light brownish-grey (10YR6/2) silt loam or loam with moderate numbers of gravel and stone fragments; weakly coarse platy; firm to very firm and moderately compact in places; large roots penetrate this horizon

(After U.S.D.A.)

Brown earths and related soils (Podzolic soils)

The brown earths are genetically related to the podzols, and in the sense that they are all leached soils, they may be described as *podzolic*. It is important to note the usage of the word podzolic, for it describes soils which are by no means fully developed podzols. These soils are evenly distributed amongst the five continents and cover about 7 per cent of the land surface of the world (Fig. 5.1). The name brown earth is translated from *Braunerde*, the name given by the German scientist Ramann to soils of a uniform brown colour, lacking the distinctive horizons of the podzol, and occurring in central Europe. Some confusion is evident in the use of names given to these soils. Brown forest soils were originally described in Europe as being developed under oak or beech forest, and their descriptions usually stress a calcareous parent material. *Grey-brown podzolic* is the name given in the U.S.A. to describe soils with a thin litter horizon and a brown blocky B horizon or illuvial clay accumulation, whilst similar soils in Russia are called *dern-podzolic soils*. Recent work by French pedologists has clarified the nomenclature for western Europe, and as their lead has found considerable support in Britain and elsewhere, this account will use their approach.

In Europe and North America the parent materials of these soils are very variable as they are frequently formed from the deposits left after the Pleistocene Glaciations. South of the extent of the glacial deposits brown earths are characteristically found on sandstones, siltstones, clays and loess.

Brown earths and related soils are associated with those areas of the world's land surface originally covered with deciduous forest. In Europe these forests are characterised by oak, beech and ash woodlands, and in America by oak, beech, maple and hickory. Although now largely cleared, these forests originally had a canopy formed by the dominant trees and a rich undergrowth of smaller trees, herbs and grasses.

The climate in which these soils develop is not as extreme as that of the podzol zone to the north, and corresponds to the Köppen types, marine west coast (Cfb) and humid coastal climates (Dfa, Dfb, Dw, Dwb). Winter temperatures average about 0 °C for less than three months of the year, and summer temperatures rarely exceed the range of 21–26 °C. Rainfall is evenly distributed throughout the year with a maximum occurring during the autumn months. Continued drought is rare. The amount of water received by the soil is sufficient to cause a moderate amount of leaching, but not enough to cause podzolization.

With the annual leaf-fall of the deciduous trees and the contribution from the smaller shrubs, herbs and grasses, the litter produced is more varied than that of the coniferous forest or heath. It is also of a higher nutritive status and is more easily digested by the soil-living fauna, which are present in greater numbers than in podzol soils. An efficient breakdown of plant tissue results. The humus is incorporated into the soil by the action of earthworms who drag material into their burrows from the surface, and ingest a mixture of mineral and organic matter. In their digestive tract, the mineral and organic matter

is brought into very close association and when defecated comes to form mull humus. Therefore, the brown earth profile frequently lacks the superficial horizons of organic material other than the current year's leaf-fall. However, forms transitional to the podzol have acid mull, or even moder, where some superficial accumulation of organic material occurs. The A horizon is enriched by the presence of organic matter to between 3 and 5 per cent by weight in woodland soils. Most brown earths are used extensively for agriculture, the original forest having long since been cleared.

A typical brown earth should possess the following salient features. It should be leached of carbonates, although some may be present in the C horizon or as added lime, and as a result it should be neutral to moderately acid (pH 4·5–6·5) with the base saturation and pH increasing with depth from the surface. The humus material should be of the mull type, although this may range from calcareous mull to acid mull or even moder in some soils transitional to the podzol group. The composition of the clay fraction should be constant throughout the profile as shown by the silica:sesquioxide ratio. The profile should be freely drained, although transitional forms to the gley group do occur.

Several sub-groups of the brown earths have been recognised which depend upon the type of humus produced and the physical nature of the parent material. The *acid brown soils* (sols bruns acides) are developed upon sandy or silty parent materials with an acid mull or moder type of humus. *Leached brown soils* (sols bruns lessivés) are formed with a mull humus on parent materials richer in clay, and are clearly transitional in form to the leached soils. Soils developed upon certain ironstone formations of the Jurassic sequence give rise to *ferritic brown earths*. Soils formed on alluvial materials, known as *brown warp soils*, occur on the better-drained positions of floodplains, and in these, as in the other types of brown earth already mentioned, evidence of gleying in the lower part of the profile has led to the recognition of related *brown earths with gleying*.

Acid brown soils (Dystrochrepts) develop on hard, base-deficient rocks such as sandstones and siltstones or coarse-grained igneous rocks (Plate 7). These soils were first recognised as a separate sub-group in the Ardennes region. They lack any illuvial horizon with an accumulation of clay, iron or dispersed humus, but have a (B) horizon which is distinguished by its structure and sometimes by its slightly brighter colour. These soils are strongly acid, pH 4·5, and have a low base content, which led to their original British name of 'brown earths of low base status'. Clay content is usually less than 20 per cent. There is a greater tendency for organic matter to form L and F horizons of the moder type in these acid soils. The mineral soil below is humus stained for 5 to 8 cm., to form the A horizon below which a transitional horizon can be distinguished. The (B) horizon frequently has a brighter reddish or orange coloration, which gradually reduces in intensity with depth, eventually the C horizon is reached at about 50 cm. (Fig. 7.7 and Table 7.2).

Profile of an acid brown soil from Carhaix, Finistère
(Parent material – fine-grained acid schists)

A	0·25 cm.	Grey-brown silt loam, stony with crumb and black structures, porous
A/(B)	25–33 cm.	Brown, slightly more clay than above, blocky structure
(B)	33–45 cm.	Slightly more ochreous, mixed pebbles of weathering schist and blocky fine earth
C	45 cm.+	Weathered schist material

(After G. Aubert)

Leached brown soils (sols bruns lessivés) in which the clay movement is not so well developed, constitute the second group of brown earth soils. Movement of clay from the upper horizons, in particular the Eb horizon, can be demonstrated by a comparison of the clay content of the Bt horizon with that of the Eb horizon. In these soils the expression $\dfrac{\text{clay \% B}}{\text{clay \% Eb}}$ is less than two. Usually the Eb horizon of these soils is not strongly developed although it can be discerned by its lighter colour (Fig. 7.8 and Plate 8). The example below from Britain gives the salient features of these soils. It can be considered as belonging to a transitional group of soils between true brown earths and the leached soils.

Profile of a leached brown soil from Shropshire, England
(Parent material – Devonian marl)

Ap	0–20 cm.	Dark reddish-brown stoneless silt loam; fine and medium sub-angular blocky structure; friable; permeable; easily penetrated by roots
Eb	20–30 cm.	Reddish-brown stoneless silt loam; medium sub-angular blocky structure; friable; permeable; easily penetrated by roots
Bt	30–75 cm.	Somewhat brighter reddish-brown stoneless silty clay loam with prismatic structure breaking to coarse angular blocky; compact; plastic, not easily penetrated by water and roots except along cracks
C	75 cm.+	Bright reddish-brown marl, often blotched with grey-green; angular blocky or platy structure; more permeable; often calcareous below about 120 cm.

(After Mackney and Burnham)

7.7 Diagrammatic profile of an acid brown soil

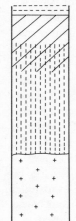

L+F		Moder or acid mull humus
A		Mixed mineral and organic horizon, strongly acid
A/ (B)		Transitional horizon, probably depleted by leaching
(B)		Weathered horizon without appreciable enrichment with colloidal material discerned by slight differences of structure and colour
C		Little altered sandstone, siltstone or glacial sand

7.8 Diagrammatic profile of a leached brown soil

L		Acid mull humus
A		Mixed mineral and organic horizon, strongly acid
Eb		Lighter-coloured eluvial horizon, strongly acid and depleted of clay
Bt		Illuvial horizon with clay enrichment moderately or strongly acid
C		Little altered parent material such as Keuper marl or boulder clay, possibly calcareous

As clay eluviation is such a distinctive feature, some pedologists are reluctant to classify the leached soils with the brown earths. The original definition of a brown earth does not include any mention of clay movement and it would seem sensible to draw the dividing line between brown earths and leached soils in a similar position to that between Inceptisols and Alfisols of the U.S.D.A. *7th Approximation.*

Table 7.2 *Sub-groups and horizon sequence of podzolic soils*

Soil Group	Soil Sub-group	Horizon Sequence		
Brown earths	Acid brown soils	A	A(B)	(B) C
	Leached brown soils	A Eb	Bt	C
Leached soils	Grey-brown podzolic soils			
	Sols lessivés	A Eb	Bt	C
	Dern-podzolic soils			
Grey soils	Grey forest soils	A Eb	Bt	C
	Grey wooded soils	A Ea	Bt	C

Leached soils

Leached soils (Alfisols) are formed from decalcified medium- or fine-textured parent materials such as are derived from the Keuper Marl, loess or glacial till. Following the removal of lime, it would appear that bases are leached from the exchange positions leading to an increase in acidity. As the depth of leached soil increases and the flocculating effect of the calcium ions is reduced, a gradual movement of clay particles begins. These are washed (lessivé) from the A and Eb horizons into the Bt horizon where the clay particles come to lie on and parallel to the ped faces. By this means the Bt horizon gradually has its clay content increased and comes to have a well-developed medium prismatic or blocky structure which contrasts with the weaker blocky structures of the Eb and A horizons. This horizon with the greater clay content is known as a textural B horizon. The clay is not broken down chemically but particles are washed down the pores and cracks in the soil, particularly after the period of summer drying when shrinkage cracks are most evident (Figs. 3.8 and 3.9).

The leached soils (sols lessivés) are those soils which have a ratio of the clay content in the B horizon compared with that in the A and Eb horizons which is greater than two, and the Eb horizon is clearly developed. Although these soils can be seen in the British Isles, the increasingly continental conditions of mainland Europe seem to encourage their development. The feature of strong clay movement is a characteristic which these soils share with the grey-brown podzolic soils (Udalfs) of the U.S.A. and the dern-podzolic soils of the U.S.S.R. (Table 7.2).

Profile of a leached soil (sol lessivé) from the Paris region
(Parent material – Loess overlying sandstone)

A	0–7 cm.	Very dark brown (7.5YR2/2) silt loam with weak crumb structure many roots
Eb	7–21 cm.	Yellowish-brown (10YR5/4) silt, moderately well-developed blocky structure; many roots; gradual boundary
Bt₁	21–65 cm.	Light yellowish-brown (10YR6/4) silt with well-developed blocky structure; common roots; irregular boundary
Bt₂	65–97 cm.	Yellowish-brown (10YR6/4) silty clay with medium and fine blocky structure well developed; brown (7.5YR4/4) clay skins (1 mm.), numerous and obvious; roots common, clear boundary
C₁	97–112 cm.	Loess in which some cracks are filled with clay
C/D	112–180 cm.	Pale brown loess overlying gritstone

(After N. Federoff)

On pervious parent materials such as loess these soils remain freely drained, but it will be appreciated that the Bt horizon with fewer cracks, and these partially infilled with clay, can present an impedance to water moving down the profile. Therefore, on clayey materials with level sites, and particularly where lateral seepage occurs, features of gleying are apparent (Plates 9 and 29). Thus the leached brown soils and the leached soils may merge laterally into analogous soils with gleying, or even surface-water gley soils depending upon the intensity of gleying conditions at any given site.

Variations such as the *rendzina* (Plate 28) *ranker* and *ferritic brown earth* occur within the brown earth zone of soil formation (Fig. 7.9). The influence of parent material plays a fairly strong part in deciding the type of soil produced. These soils will be described more fully in the chapter on intrazonal and azonal soils.

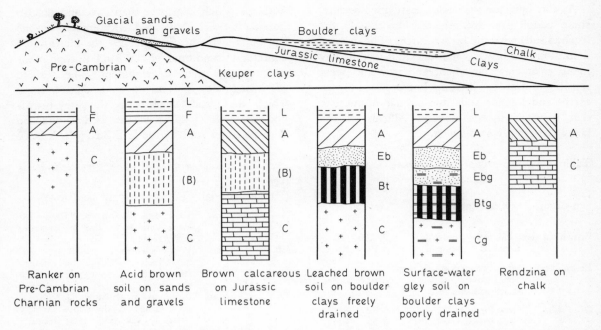

Ranker on Pre-Cambrian Charnian rocks

Acid brown soil on sands and gravels

Brown calcareous on Jurassic limestone

Leached brown soil on boulder clays freely drained

Surface-water gley soil on boulder clays poorly drained

Rendzina on chalk

7.9 Diagrammatic section of Lowland Britain from near Leicester to the North Sea indicating relationship between soils and parent materials

Grey soils

Grey forest soils and grey wooded soils (Boralfs), formed in a more continental environment, are found between the podzol zone to the north and the chernozem zones of soil formation in North America and in Russia (Fig. 5.1). The climate in which these soils have been formed has mean annual temperatures of −9 °C in January and 19 °C in July. Precipitation is light, 550 mm. per annum, 45 per cent of which falls as early summer rain. These are average figures for European Russia. There is a variation from east to west with both rainfall and length of frost-free period increasing westwards. Similar climatic conditions exist in the North American districts where these soils occur. Climatologically, this zone is described by Thornthwaite as 'subhumid and semi-arid microthermal climate' and by Köppen as 'humid continental with short summers' (Dfb, Dwb). Conditions suitable for the development of these soils do not occur in the southern hemisphere.

The parent materials of grey wooded soils in Alberta and Montana are calcareous tills, lacustrine and outwash material of Late Glacial Age.

Similar parent materials occur in U.S.S.R. such as loess-like loams and glacial deposits upon which the grey forest soils are developed.

Grey forest soils are associated with the transition from forest to steppeland. The forest of these regions has been cleared from extensive areas, but originally it was a broad-leaf forest of oak, lime, maple, birch and hazel in Europe, and poplar, spruce, fir, larch and ponderosa pine in North America. On the forest floor was a grassy herbaceous cover. Litter from this diverse plant cover has been incorporated into the soil in a mull form of humus.

The results of the soil-forming processes reflect some features of neighbouring zones. Leaching occurs to a lesser extent than it does in the brown earth soils, while evidence of calcification in the B horizon resembles that of the chernozems. The amount and the form of the humus of the grey forest soils resembles that of the leached chernozems or prairie soils. A feature of the grey forest soils, already seen in the leached soils, is the removal of clay from the upper horizons to form a Bt horizon with clay skins, indicating a process which is absent in the true chernozem soil. The

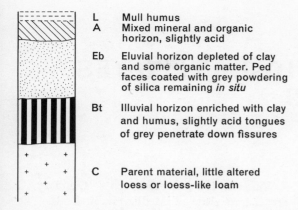

L	Mull humus
A	Mixed mineral and organic horizon, slightly acid
Eb	Eluvial horizon depleted of clay and some organic matter. Ped faces coated with grey powdering of silica remaining *in situ*
Bt	Illuvial horizon enriched with clay and humus, slightly acid tongues of grey penetrate down fissures
C	Parent material, little altered loess or loess-like loam

7.10 Diagrammatic profile of a grey forest soil

Profile of a grey forest soil from near Tula, U.S.S.R. (Parent material – non-calcareous loess overlying morainic material)

A	0–15 cm.	Dark grey clayey soil with fine nutty structure; friable; compact root mat; even boundary
A/Eb	15–40 cm.	Dark grey clayey soil (lighter when dry) with medium to coarse nutty structure; fewer roots; organic material coating peds, boundary tongues into horizon below
Eb	40–60 cm.	Brownish-grey clayey soil with pallid powdering on ped faces; prismatic-nutty structure; compact wavy boundary
Bt	60–110 cm.	Greyish-brown clayey soil with prismatic structure, more compact than horizon above; bright coatings upon ped faces
Bt/C	110–190 cm.	Brown to bright brown clayey soil with blocky to prismatic structures; coatings upon ped faces; uneven boundary. Reddish-brown sandy loam morainic material

(After V. M. Fridland)

grey colour of the A horizon is caused by the powdering of fine grains of silica remaining on the ped faces, following removal of clay to the lower horizons (Fig. 7.10 and Plate 10).

It would seem that the soils of the grey forest soil group have developed under changing biotic conditions. Profile evidence from the region of Kursk indicates that these soils may have developed under steppe conditions. Subsequently they became wooded and leaching resulted from a slight climatic change which encouraged tree growth. The presence of krotovinas, infilled burrows of steppe-living animals, in the grey forest soils as well as in the chernozems, seems to substantiate this view. The description of a representative profile from near Tula illustrates the features of these soils.

The profile of grey wooded soils resembles more closely that of the podzol with a thin raw humus horizon (5 cm.). The A horizon is not always well-developed and a pronounced platy Ea horizon tongues into the Bt horizon which has well-developed clay skins and coatings of organic matter. Although these soils have the superficial morphology of a podzol they are in fact more closely related to the grey-brown podzolic soils.

From their geographical position and the previous comments it will be appreciated that these grey forest soils and grey wooded soils are forming in areas of gradual transition. As a result their distribution will be discontinuous depending upon the interplay of the factors of soil formation. (Table 7.2).

8 SOILS OF THE MID-LATITUDES, WARM CLIMATES

Soil development in the mid-latitudes is conditioned by a wide range of climates with many transitional zones and variations. However, mid-latitude warm climates can be divided into four main categories allowing soil formation to be considered within these as a framework. They are the 'Mediterranean' areas, humid sub-tropical or southeast coastal margins, temperate continental areas and the deserts.

Soils of the Mediterranean areas

In those parts of the world which experience mild, moist winters and warm, dry summers, the vegetation and soil response is typified by that in the Mediterranean Basin. Similar environments occurring in California, southern and western Australia, South Africa and Chile are classified by Köppen as Cs climates. The soils formed are brown earths, brown and red Mediterranean soils, and cinnamon soils (Fig. 5.1).

The parent materials from which these soils are formed in Europe consist of sandstones, shales and particularly limestones and calcareous marls. The latter two are common south of the Alps. As considerable erosion has occurred there has been redistribution of the parent materials leaving rocky hills with thick colluvial accumulations in the valleys and hollows of the landscape. This redistribution is common in older landscapes, particularly in semi-arid areas such as Australia, which have been land areas for long periods of geological history. This leads to a complicated soil pattern with old soils lying adjacent to immature ones upon the same landscape. In southern Europe, northern Africa and parts of Australia, deep red clays known as terra rossa have accumulated from the weathering of the limestones of Karst areas to form a distinctive parent material for soil formation.

Winter temperatures between 5 °C and 15 °C are characteristic of Mediterranean lands. The rainfall of about 500 mm. per annum is derived from depressions which develop over the sea during the cooler season. The soils are thoroughly moistened by this rain, only to be parched again during the ensuing summer which is characterised by little or no rainfall and temperatures between 25 °C and 30 °C. The length of the summer drought appears to be an important factor in the formation of soils in this region. Increasing length of drought produces a sequence of soils ranging from brown earths developing in a leaching environment with less than one month summer drought, to the cinnamon soils where five or six months' drought leads to a calcification type of soil formation transitional to that of the continental interiors. Between these two extremes the red and brown Mediterranean soils are formed.

Summer drought

1 month   ←————————————→   5–6 months
(leaching)   (slight ferrallitization)   (calcification)
Brown earths   Brown Mediterranean soils   Cinnamon
                       ↓   soils
                 Erosion
                   ↓
Red Mediterranean soils
and soils on terra rossa

The vegetation of the Mediterranean lands originally appears to have been an evergreen forest of broadleaved and coniferous trees, in particular various species of oak and pines. Many years of human intervention have removed much of this forest by lumbering, burning and grazing. A secondary growth known as *maquis* on non-calcareous soils and *garrigue* on calcareous soils has developed as a sub-climax vegetation. Removal of the original forest initiated soil erosion over large areas.

A Mull humus
Eb Eluvial horizon which has lost clay. Slightly acid

Bt/ir Illuvial horizons which are
Bir/C enriched with clay and iron and have become red-coloured

C Limestone or calcareous marl

8.1 Diagrammatic profile of a brown Mediterranean soil

Brown Mediterranean soils (Ustalfs) developed beneath this mixed forest. They are characterised by a brown colour, a friable humus-rich A horizon which overlies a denser and less friable B horizon (Fig. 8.1). Where these soils have developed on calcareous parent materials, the upper horizons have been decalcified and clay movement has occurred. In the lower part of the illuvial horizon, and on the linings of fissures in the C horizon, clay has been redeposited in calcareous conditions as the soil dries during the hot, dry summer. At the same time siliceous-iron complexes are irreversibly precipitated to give the rich red colour of the lower horizons of these soils. This is a process of slight ferrallitization (*rubefaction*) which is even more important in tropical soils which have formed under a climate with a pronounced dry season.

Profile of a brown Mediterranean soil from Rivier Sonder End, Cape Province, South Africa
(Parent material – schists)

Ap 0–13 cm. Greyish-brown, crumbly to slightly loose gravelly sandy clay loam mixed with quartz grit and small angular shale fragments, humus deficient, but common roots in upper 10 cm. Material from B horizon has been incorporated by cultivation

B 13–63 cm. Reddish-brown gravelly clay; crumbly when dry and fairly compact when wet; stones and gravel consist of hard angular shale fragments

C_1 63–79 cm. Yellowish-brown, mottled reddish-brown partly weathered hard and soft shale with an appreciable amount of clay

C_2 79–150 cm. Slightly weathered greyish-brown shales, moderately soft

(After Van der Merwe)

Some of the red Mediterranean soils (Rhodustalfs) have resulted from soil formation in the eroded remains of these brown Mediterranean soils (Fig. 8.2 and Plate 11). Other red soils are

8.2 Mediterranean soils and relationship to landscape

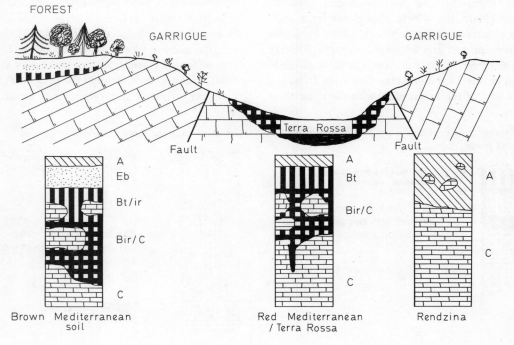

FOREST

GARRIGUE

GARRIGUE

Terra Rossa

Fault

Fault

A
Eb
Bt/ir
Bir/C
C

Brown Mediterranean soil

A
Bt
Bir/C
C

Red Mediterranean / Terra Rossa

A
C

Rendzina

formed upon the relic clays resulting from the weathering of the limestone. Often these soils occupy discontinuous pockets surrounded by rocky limestone outcrops. These soils are usually less than 1 m. in depth and although derived from limestone by solution, are often slightly calcareous through enrichment by calcium-rich solutions from surrounding areas of limestone. Their profile usually comprises a dark red clay, somewhat enriched with organic matter, which has a friable consistence. Lower horizons are formed of a firm blocky clay which becomes very plastic when wet.

Profile of a red Mediterranean soil from Tlemcen, Algeria
(Parent material – Jurassic Limestone)

A	0–20 cm.	Brown-red loam with coarse blocky structure, relatively hard consistence with numerous roots
B_1	20–50 cm.	Dark red clay with blocky structure, plastic consistence and numerous roots
B_2	50–80 cm.	Dark red clay with blocky structure, plastic consistence, common roots
C	80–160 cm.	Red clay loam with blocky structure, ped surfaces glossy, plastic consistence, numerous calcareous nodules
	160 cm.+	Limestone

(After J. H. Durand)

Red-brown earths (Ustalfs) are described from the Barossa Valley of South Australia, an area which experiences a Mediterranean type of climate (Plate 12). These soils occur on a wide range of parent materials which are to be found between the 350 and 630 mm. isohyets. As they are subjected to strong weathering, these red-brown earths have some clay eluviation from the upper part of the soil profile, with the deposition of clay and calcium carbonate in the lower horizons. Movement of these constituents is dependent upon the seasonal rainfall which causes periodic saturation and downward percolation (Fig. 8.3).

Profile of a red-brown earth, Barossa Valley, South Australia
(Parent material – Pleistocene colluvial deposits)

A_1	0–8 cm.	Light brown hard, compact and cemented loam
A_2	8–15 cm.	Light reddish-brown compact clay loam
B_1	15–60 cm.	Dark reddish-brown friable granular to granular nutty clay which becomes blocky in the lower part of the horizon
B_{ca}	60–135 cm.	Dark reddish-brown clay with small amounts of lime
C	135 cm.+	Mottled brown and reddish-brown clay continues to 300 cm.

(After C.S.I.R.O.)

Non-calcic brown soils are rather similar to the red-brown earths, but they lack the calcareous B_{ca} horizons. Instead, the B horizon of the non-calcic brown soil can be distinguished by a prismatic structure and a change in texture. These soils are characteristic of the wheat belt of Western Australia.

Where a longer summer drought is experienced, the characteristics of the soil gradually become more typical of the drier, semi-arid steppelands. A light yellowish-brown soil results which Russian authors call a *cinnamon* soil (Ustrochrept). The general features of this soil are that it has a blocky structure and a clay content which increases from the lower part of the A horizon downwards through the profile (Fig. 8.4). Calcium carbonate concretions may occur below

8.3 Diagrammatic profile of a red Mediterranean soil and terra rossa

A	Thin mull humus forming in eroded remains
Bt	of B horizon
Bir/C	Illuvial horizons which are enriched with clay and iron and have become red-coloured
C	Limestone or calcareous marls

8.4 Diagrammatic profile of a cinnamon soil

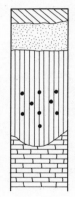

A	Mull humus. Mixed organic mineral horizon, neutral or slightly acid
Eb	
Bca	Undifferentiated B horizon, increasing clay content with depth, calcareous
C	Slightly calcareous clay parent material

about 30 cm. in these soils in drier localities. The A horizon is reported to be moderately rich in organic matter, ranging between 4 and 7 per cent. These cinnamon soils occur in eastern Spain, the Balkans, Turkey, North Africa, as well as in California, Mexico, Chile, China and Australia. They are the zonal type which has developed a response to the drier type of Mediterranean climate.

Profile of a cinnamon brown soil from Khasavyurta, Daghestan
(Parent material – not given)

A1	0–10 cm.	Cinnamon-brown with dark grey cast, silty-blocky, mellow. Matted with roots. Does not effervesce. Gradual transition
A2	10–41 cm.	Darker colour than above, prismoidal-blocky, somewhat compact. Does not effervesce. Gradual transition
B1	41–55 cm.	Grey with a brown tone, prismoidal-blocky, somewhat compact. Contains pseudo-mycelium (p. 63). Fine clay loam, effervesces. Distinct transition
B2	55–75 cm.	Dark pale-yellow with white spots. Blocky-lumpy compact contains large quantities of pseudo-mycelium. Fine clay loam effervesces vigorously to the end of the profile
C1	75–118 cm.	Lighter colour than above, cloddy, compact, efflorescences of carbonate medium clay loam
C2	118–170 cm.	Dark pale-yellow, cloddy, compact, carbonate in the form of spots and veins. Clay loam
C3	170–265 cm.	Dark pale-yellow, lumpy, compact, coarse clay loam

(After Zalibekov)

Various intrazonal soils also occur which deserve mention. On the many outcrops of limestone rocks in the Mediterranean Basin, rendzinas (rendolls) occur with fine-textured, crumb-structured, calcareous shallow profiles. Similar shallow soils on clays and other non-calcareous parent material are termed *rankers* (lithic hapludents). These are often associated with the higher and steeper mountainous areas. Poor drainage is not a characteristic generally associated with the soils of these regions, for although examples of gley soils do occur, most soils, even river alluviums are not gleyed. Some alluvial soils may show signs of salt enrichment, particularly where salty waters have been used for irrigation in areas of less than 500 mm. annual rainfall. Seasonally poorly drained soils similar to those of the tropical regions are reported from Morocco and other parts of North Africa. These soils, called *tirs* (usterts or xererts), are black or dark brown fine-textured clays which occupy the lower parts of the landscape. Soils with crusts of calcium carbonate (*croûte calcaire*) have been called *kafkalla* in Cyprus. Similar soils also occur in South Australia (Plate 13). These crusts are concretionary horizons formed by an accumulation of calcium carbonate within the soil profile which is subsequently revealed at the surface by erosion.

Soils of the warm-temperate east margin climate

Occupying a similar position to the Mediterranean climates but on the eastern sides of the continental land masses are those areas which experience the warm-temperate east margin type of climate. The zonal soil type of this climate is that exemplified by the red and yellow podzolic soil. Features observed in these soils suggest the effects of podzolization, whilst other features are suggestive of ferrallitization. Because geographically these soils occur midway between the main areas of podzolization and ferrallitization, it is not surprising to find evidence of both processes present. The main areas where these soils can be seen are in the southeast of the United States, central China, eastern Australia and eastern Brazil (Fig. 5.1).

The climatic regime in which the red and yellow podzolic soils are formed is one of hot, humid summers with convective rainfall, and short mild winters with precipitation from frontal activity between air masses of different character. Rainfall is well distributed throughout the year, and amounts to between 1250 and 1500 mm. per annum. Average temperatures range from around 5 °C in the coldest month to about 25 °C in the summer. This climate is called humid sub-tropical (Cfa) by Köppen.

Granites, gneisses, schists, sandstones, shales, limestones and various unconsolidated sediments are quoted as parent materials for these soils in the United States. Most of these parent materials are well-weathered, siliceous and situated upon geomorphologically old land surfaces.

The deciduous forests which remain indicate a former cover composed of oak, hickory and chestnut on the freely drained upland areas. Low-lying, poorly-drained land carried stands of slash and loblolly pine with cypress. Whilst these trees formed the upper storey, they allowed sufficient light to penetrate through to the forest floor for a dense shrub and herbaceous under-storey to develop.

Red-yellow podzolic soils are defined as 'a group of well-developed, well-drained acid soils having thin organic and organic-mineral horizons over a light-coloured bleached horizon, over a red, yellowish-red or yellow, more clayey B horizon'.

The moist climate throughout the year is conducive to leaching, and the evidence for this is seen in the acid surface horizons. A maximum amount of clay occurs deeper in the profile, but this seems to be caused for the most part by kaolinite formation *in situ* rather than by trans-location of clay down the profile. The red and yellow colours of these soils indicate different degrees of hydration of the iron oxides; the red soils develop in drier conditions and the yellow where moister conditions prevail (Fig. 8.5 and Plate 14).

Profile of a red podzolic soil from Virginia, U.S.A.
(Parent material – weathered gneiss)

A₀₀		Thin layer of leaves and pine needles
A₁	0–5 cm.	Brownish-grey very friable sandy loam with fine weak crumb structure, strongly acid
A₂	5–20 cm.	Weak yellow to light yellowish-brown nearly loose or very friable sandy loam, strongly acid
B₁	20–25 cm.	Weak reddish-brown to strong brown friable sandy loam or light sandy clay loam with medium granular structure, strongly acid
B₂	25–95 cm.	Moderate to strong reddish-brown clay, plastic when wet, very firm when moist and very hard when dry. Medium blocky structure. Some white sand grains and small mica flakes, strongly acid
B₃	95–150 cm.	Light to moderate reddish-brown clay loam with mottles of yellow, firm to friable when moist. Weak coarse blocky structure with enough mica flakes to make it feel slippery, strongly acid
C	150 cm.+	Mottled light reddish-brown, yellowish-brown, light grey and black friable disintegrated rock material

(After U.S.A.)

L+F	
A	Acid mineral and organic horizon
Eb	Lighter coloured acid eluvial horizon thicker in yellow soils
B	Red or brownish-red horizon rich in clay which increases with depth. Iron more hydrated in yellow soils
C	Parent material formed from wide range of geological materials

8.5 Diagrammatic profile of a red-yellow podzolic soil

Profile of a yellow podzolic soil from Cheltenham, near Sydney, Australia
(Parent material – sandy shale)

A₁	0–5 cm.	Grey to dark-brown (10YR5/1, 3/3), friable, crumb-structured loam or silt loam
A₂	5–25 cm.	Very pale brown to yellowish-brown (10YR7/4, 5/4) friable loam without distinct structure
B₂	25–50 cm.	Yellowish-brown (7.5YR5/6) fine blocky fairly stiff clay with small pieces of ferruginous very fine-grained sandstone. Friable when moist and plastic when wet
C	50 cm.+	Slabby, very fine-grained sandstone with streaks of brown iron oxide

(After Thorp)

In the past, American pedologists thought a change in climate had occurred since these soils were originally formed, so that the present leached profile had developed from a soil which had more affinities with those in the more humid tropics. If this were so, the red and yellow podzolic soils would resemble the red and brown Mediterranean soils, which are thought by French pedologists to possess signs of leaching as well as ferrallitization. Recent American work has tended to discredit climatic change as an important factor in the formation of the red and yellow podzolic soils.

Although the southeast of the U.S.A. is where most work has been done upon their development, these soils are not confined to this region. Similar soils are described from the U.S.S.R. in the state of Georgia where they occur on the shores of the Black and Caspian Seas, south of the main Caucasian ranges. In Australia, particularly in New South Wales, areas which experience 500 to 650 mm. rainfall per annum develop these soils. They occur extensively upon the elevated tablelands in the eastern part of the state, where they are developed upon Palaeozoic sediments. The red podzolic soils occur upon the freely drained interfluve sites, and the yellow podzolic soils occupy the lower sites in the landscape. In Australia with less rainfall, these yellow podzolic soils are frequently affected by the presence of salts, leached from upslope, and yet not transported right out of the landscape. As a result, features of a solonetzic nature can be seen in the profiles.

Although widely known by the name red-yellow podzolic soils, they have recently been reclassified in the *7th Approximation* as belonging to the order Ultisols, sub-order Udults. However, as research is still in progress, further redefinition is likely at great soil group level.

Soils of the temperate continental interiors

The *chernozems*, or *black earths*, occur on the natural grasslands of the North American prairie and Russian steppes. The formation of these soils results from a delicate balance of temperature, rainfall and vegetation brought about by their geographical location in the continental interiors. While the chestnut soils are developed towards the drier desert margins, prairie soils and varieties of leached chernozem form a transition to the leached soils of cool temperate climates.

Chernozems

In parts of Romania, Hungary and the Ukraine southeast of Kiev, as well as west-central Siberia, the main soil type is the chernozem. This soil also occurs in North America, west of longitude 95° W and north of the Arkansas River, but the area is very much smaller than that of the Eurasian chernozems. According to Russian sources, almost 2,000,000 square kilometres of their country is covered by chernozems, which is about half the world area of these soils. Chernozems, or black earths similar to them, occur in all habitable continents, but there are differences of opinion about the status of black earth soils in Africa (*vlei* soils), Australia, India (regur or black cotton soil) and South America which do not possess all the properties of the true chernozem (see Vertisols, pp. 70–1).

The chernozems (Borolls) have developed south of the forest-steppe ecotone on the steppes of Russia and the prairies of U.S.A. and Canada. In both the continents where they are of major occurrence, the chernozems are developed upon loess or loess-like parent materials. Outside those

8.6 Distribution of loess throughout the world (after Lobeck)

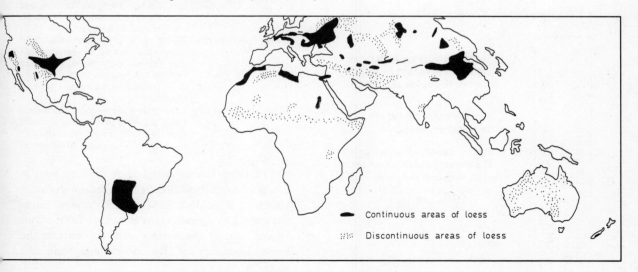

● Continuous areas of loess

⠿ Discontinuous areas of loess

areas of the northern hemisphere covered by ice during the Pleistocene Glaciations, tundra climate prevailed, and as the glaciers melted much glacial debris from moraines and outwash streams was left without a vegetative cover. Winnowed by the wind, the fine dust from these deposits was transported and redeposited as loess. This has formed the parent material for the chernozem soils in favourable areas, though it is also found as a constituent of the parent material of soils elsewhere (Fig. 8.6).

The climate of the continental interior has a cold winter with temperatures between -7 °C and -10 °C, and the soil is frozen to a depth of 60–80 cm. from November to April. The winter snow cover is not deep. There is an annual precipitation of 550 mm., with a slight rainfall maximum in the summer months. Temperatures in July are in the range 19 °C–21 °C. There is a period of 150 to 160 days without a killing frost during the year. Chernozem soils occur in a number of climatic divisions. According to Köppen these are humid continental with short summer (Cwa), humid continental with long summer (Dfa, Dwa), and semi-arid, middle-latitude steppe (BSk).

The chernozem soils occur naturally beneath a grassland composed of a large number of genera including *Agropyron*, *Bouteloua*, *Buchloe*, *Poa* and *Stipa*. Herbaceous plants are common, and patches of trees including oak and lime are characteristic of the ecotone of the forests to the north. This landscape, a gently undulating plain where grasses and small herbaceous plants once formed a dense ground cover, is now mainly cultivated.

The environmental factors which control the formation of these soils depend on warm spring and summer temperatures with adequate moisture supply from snow-melt and from the early summer rain. The rapid growth of grasses and herbs produces a large amount of root and aerial shoots. The drought of late summer and the frosts of winter largely arrest the process of decomposition. Consequently, losses of organic matter are minimised, and as humus formation takes place in a neutral environment rich in calcium, the mull or calcareous mull form of humus results. Chernozems have a rich fauna which incorporate the humus into the deep A horizon, and the soil is also worked through by small vertebrates, the former presence of which is shown by their infilled burrows, known as krotovinas (Fig. 8.7 and Plate 15).

Profile of a typical chernozem from near Kursk, U.S.S.R.
(Parent material – loess over sandstone)

	0–5 cm.	Compact root mat of grasses
A_{11}	5–60 cm.	Uniform dark grey clayey upper part of the humus horizon with fine crumb-granular structure; friable; roots mostly in the 0–40 cm. layer; gradual boundary
A_{12}	60–100 cm.	Dark grey, cinnamonic-tinged, clayey lower part of the humus horizon with crumb-nutty structure; more compact consistence than above; effervescence below 90 cm. gradual boundary
A/C_{ca}	100–120 cm.	Many dark grey krotovinas and worm burrows in a cinnamonic-pale yellow background; silty clay with crumb-nutty structure; abundant pseudo-mycelium; gradual boundary
C_{ca}	120–250 cm.	Light pale-yellow loamy calcareous horizon; weak prismatic structure; firm; porous; dispersed carbonate

(After Afanasyeva)

The deep humus-enriched A horizon 80–100 cm. thick with its well-developed crumb-granular structure is characteristic of the 'typical chernozem'. The humus content ranges from about 10 per cent at the surface to 2 per cent at the lower boundary of the A horizon, and the

8.7 Diagrammatic profile of a chernozem

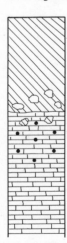

A Mull humus incorporated to considerable depth by earthworms, neutral or slightly acid

Krotovinas (burrows) of vertebrate animals

Cca Parent material of loess or loess-like loams, concentration of $CaCO_3$ in Cca horizon but depth varies according to amount of leaching
C

carbon:nitrogen ratio is in the range 11–12. As the clay minerals are of the montmorillonitic variety, the exchange capacity and fertility is high. Chernozems are slightly leached, so their upper horizons are neutral or slightly acid. The passage of moisture through the profile is downwards in spring following snow-melt. During summer evaporation from the surface reverses the process so that the soil is rarely wetted to beyond a depth of 1·5 to 2·0 m. It is this mild leaching and re-evaporation which leads to the concentration of calcareous material in the lower part of the profile and which retains soil nutrients within the rooting zone.

The calcium carbonate horizon in the example given extends from 90 to 180 cm., and has a maximum content at 120–130 cm. The upper part of the accumulation is characterised by '*pseudo-mycelia*', a filamentous form of carbonate concretion. In the zone of maximum accumulation carbonate concretions occupy the former pores and cavities in the loess, and even form small nodules. The typical chernozem has a maximum carbonate content in the upper part of the ca horizon, which occurs at shallow depth, even in the A horizon. Therefore, it can be distinguished from the leached chernozem where the accumulation occurs lower in the profile and the maximum carbonate content occurs in the middle of the ca horizon. These variations of leaching can be related to the micro-relief, for with the spring snow-melt, water is concentrated into the depressions of the landscape where more leaching takes place than on higher areas (Fig. 8.8).

In North America, the prairie soils (Udolls), are roughly equivalent to these degraded (leached) chernozems. Both occur on the moister, wooded boundary of the typical chernozems. Prairie soils have a profile in which a dark, humus-rich A horizon overlies a brown, compact B horizon. This B horizon is absent from typical chernozems and represents the first sign of an illuvial horizon in the sequence of soils from chernozem to grey-brown podzolic soils. The horizon of calcium carbonate accumulation occurs below the B horizon, which effectively separates it from the organic-rich surface horizon. The prairie soils with their leached profile typical of forested soils further north and with their organic-rich horizons typical of the grasslands, present problems of

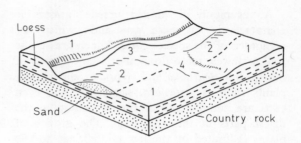

8.8 Chernozems and their relationship to the landscape
1. Typical (deep) chernozems developed on freely-drained interfluve areas
2. Sandy river terraces with leached brown earth or even podzol soils
3. Soils of the alluvium
4. Leached and podzolized chernozems of the topographic depressions

genesis which are as yet unsolved. Towards regions of drier climates, the prairie soils become redder in colour and are known as reddish-brown prairie soils.

In regions with a drier climate the production of organic matter for the A horizon becomes less, leading Russian pedologists to recognise *ordinary* and *southern* chernozems. Ordinary chernozems have less organic matter (5·0 per cent) and southern chernozems an even lower amount (3·50 per cent). Higher temperatures cause the oxidation of the organic matter to proceed at a greater rate, while with a lower rainfall the volume of plant material produced is less. The thickness of the A horizon decreases, and the zone of carbonate accumulation is nearer the surface (Plate 16).

Chestnut soils

Both American and Russian pedologists have described soils of the drier parts of the short-grass steppe and prairie as *chestnut soils* (Ustolls) (Plate 17). These soils occur in the extreme south of the Ukraine, and in a broad arc from the western shore of the Caspian Sea eastwards along latitude 50 °N as far as the Irtysh River. Beyond the Altai Mountains these soils are widespread in eastern Outer Mongolia and northern Manchuria. In North America, chestnut soils extend from the North Saskatchewan River south-eastwards to the Llano Estacado on the High Plains east of the Rockies.

A vegetation of 'mixed prairie' grasses is characteristic of these grasslands which are composed of the species *Stipa*, bunch grasses, as well as lower, more drought-resistant grasses of the species *Bouteloua* and *Aristida*. Salt-tolerant plants such as *Artemisia*, sagebrush, and even cacti are components of the vegetation in the driest parts of this soil zone. The growth of plants is limited by the low rainfall, 340 to 360 mm. per annum, and there is a high rate of evapotranspiration. Temperatures of 20 °C to 25 °C are experienced during the summer, and freezing winter temperatures are possible over much of the area where these soils occur. The supply of organic matter is much less than in the chernozem soils, and the humus is of the mull type. The A horizon is reduced in thickness to less than 25 cm. and the horizon of calcium carbonate accumulation occurs between 40 cm. and 50 cm. from the surface.

The parent material of these soils in the U.S.S.R. is mostly a loess overlying older sediments, but the example given from U.S.A. is formed from a calcareous drift.

Profile of a chestnut soil from Williams County, North Dakota
(Parent material – calcareous glacial till)

A_{11} 0–3 cm. Brown (10YR5/2 dry), to very dark brown (10YR2/2 moist) loam with soft crumb structure; neutral or mildly alkaline

A_{12} 3–10 cm. Brown (10YR4/2 dry), to very dark brown (10YR2/2 moist) loam with weak platy structure readily crushed to a medium crumb. Neutral or mildly alkaline

A_3 10–25 cm. Brown (10YR4/2 dry), to dark brown (10YR3/2 moist) silt loam with moderate prismatic structure. Neutral to slightly acid

B_2 25–50 cm. Greyish-brown (2.5Y5/2 dry) to dark greyish-brown (2.5Y4/2 moist) heavy loam with strong prismatic structure in the upper part and very coarse blocky below; slightly calcareous

C_{ca} 50–65 cm. Light greyish-brown (2.5Y6/2 dry) to yellowish-brown (2.5Y5/4 moist) friable massive or weak coarse subangular blocky loam or silt loam, highly calcareous

C 65 cm.+ Light grey (2.5Y7/2 dry) to yellowish-brown (2.5Y5/4 moist) sandy clay or clay loam till

(After U.S.D.A.)

An interesting pattern of soil distribution has developed on the drier steppes of the southern Ukraine in the zone of chestnut soils. This pattern is related to the micro-relief. Broad shallow depressions occur which receive water following the spring snow-melt. Because of the movement of water through the soil into these depressions, there is a tendency for soluble salts to accumulate, but at the same time, the additional water can cause leaching of the salts to greater depth in the soils of the depressions (Fig. 8.9). A dynamic situation exists with the salts migrating down through the soil in a wetter season, and accumulating nearer the surface in a drier season. This has led to the formation of a whole range of solonetz and solodized solonetz soils where halomorphic and hydromorphic features are associated (Chapter 10).

Soils of the desert

It is estimated by Russian pedologists that 17 per cent of the earth's surface has desert soils. Their extent varies considerably from continent to continent, with Australia 44 per cent, Africa 37 per cent and Eurasia 15 per cent of their land area coming in to this category. The desert regions of the world have a severe climate, in which special weathering conditions prevail and in which there is a highly specialised plant and animal life. Almost all soil classifications distinguish a group of desert soils. These occur in a broad zone across Africa and Asia where they are interrupted by the various mountain ranges, the central part of Australia, and smaller areas in North and South America.

The climate of the deserts is characterised by an irregular and insufficient rainfall which does not provide enough moisture for leaching. Long periods of complete drought may be broken by brief torrential showers, while the average rainfall may be less than 150 mm. per annum. Temperatures during the day are high, but during the night there is a rapid fall in temperature, often ranging over 40 °C. Frost is common on more elevated areas of the temperate deserts. Low atmospheric humidity and few clouds combine to give uninterrupted sunshine, so that soil-surface temperatures of 43 °C are recorded. These areas are classified by Köppen as middle latitude desert (Bwk) and low latitude desert (Bwh).

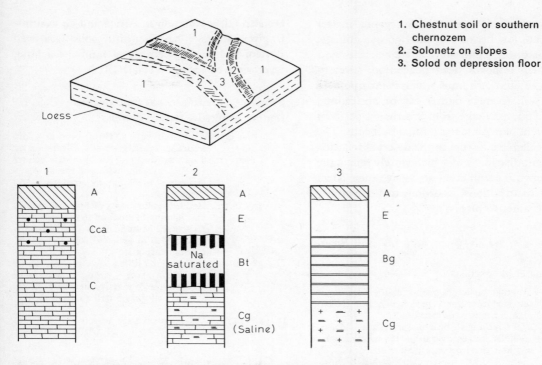

1. Chestnut soil or southern chernozem
2. Solonetz on slopes
3. Solod on depression floor

8.9 Soils of the chestnut and southern chernozem zone of the southern Ukraine

Above ground-level mechanical weathering produces a coarse regolith, and although there is a lack of surface water in the deserts, some chemical weathering occurs below ground. The weathered material is redistributed by wind and sheetflood, so that the parent material for soil formation is often well differentiated into rocky areas, dune areas with sandy textures and playa floors with silt or clay textures. Salts tend to be accumulated when waters evaporate from the lower-lying areas (Fig. 8.10). Although the presence of salts is widespread in arid areas, they do not occur in all desert soils; usually it is a shortage of water rather than the presence of salts which is the limiting factor for plant life. Plant nutrients may be available in suitable forms, but the lack of water inhibits their uptake and the growth of plants is retarded.

Even though there is insufficient moisture for downward leaching, there are movements of soil constituents in desert soils. Some movement occurs in an upward direction as is shown by the crusts of calcium carbonate and calcium sulphate, as well as the *desert lac* of iron oxides. Whilst

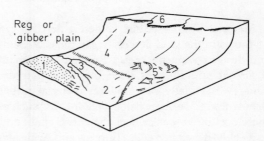

8.10 Block diagram to show the relationship of desert soils to the landscape
1. Reg or stony desert
2. Clay pan (saline)
3. Alluvial tract
4. Red or grey desert soils
5. Erg or dune sand
6. Plateau formed by duricrust

some of these may be contemporary features, others have been formed in a previously moister climate of a Pleistocene pluvial period and are thus relic features. Illuvial accumulations can be revealed at the surface by the winnowing away of the overlying soil material.

Plant growth is sparse and organic matter supplies are low because of the rigorous climate. Strong oxidising conditions at the surface, and wind which removes dead vegetation, severely limit the amount of humus which is incorporated into the soil. Average figures can be misleading, but those quoted range from less than $\frac{1}{2}$ per cent to 2 per cent of organic matter in desert soils. The lack of available moisture in the desert is reflected in the plant distributions which closely reflect the soil patterns. The different soils have their effect on plant distributions according to their ability to supply water to plants.

Profile of a desert soil from near the Sak River, South Africa

(Parent material – schists)

Surface	Angular shale fragments and stones cover the surface (desert pavement), but desert pigment is not conspicuous
0–15 cm.	Light yellowish-brown sandy clay loam; vesicular, and easily crushed to powder; roots few and thin; organic matter absent
15–45 cm.	Light reddish-brown with whitish tint, coarse sandy clay loam; fairly dense; occasional calcium carbonate concretion
45 cm. +	Platy undecomposed schist

(After C. R. Van der Merwe)

Material best described as desert detritus can be found in the sand deserts (*ergs*), the clay plains and desert pavements (*regs*). Profile development is absent and there is little biological activity. This is essentially raw mineral material rather than true soil. In Africa alone, 'non-soils' such as these occupy 28 per cent of the entire continent.

Weakly-developed soils (Entisols and Aridisols) are also fairly common in Africa and an example is given above. These youthful soils lack well developed horizons because of limited leaching, or erosion, or accretion (Plate 18).

Profile of a grey desert soil from Colorado, U.S.A.

(Parent material – terrace gravels)

A_1	0–1 cm.	Light brown (7.5YR6/4) loam in a soft vesicular crust, generally calcareous
A_2	1–10 cm.	Light brown (7.5YR6/4) to pale brown (10YR7/4) loam with soft, weak platy structure breaking to granular, calcareous
B_2	10–38 cm.	Reddish-yellow (7.5YR6/6) calcareous clay loam with medium to coarse blocky structure. Mottled with pinkish-white (7.5YR8/2) in areas of soft segregated calcium carbonate
C_{ca}	38–80 cm.	White (10YR8/2) to pinkish-white (7.5YR8/2), very strongly calcareous clay loam with massive structure. Lime occurs well distributed throughout the soil mass
D_{ca}	80 cm. +	Calcareous terrace gravels

(After U.S.D.A.)

Grey desert soils or *sierozem* occur in areas where there is about 250 mm. or less average annual rainfall which comes in irregular showers. With a limited amount of organic matter, these soils have free calcium carbonate at or just below the surface. They are developed beneath a vegetation described as 'desert-shrub' in the intermontane valleys of Colorado, New Mexico and Utah. The sagebrush (*Artemisia*) and bunch-grasses are the most common components of this vegetation. In Russia the grey desert soils occur in the foothill regions of Turkestan (southern Kazakh S.S.R.).

9 SOILS OF THE LOW LATITUDES

The discussion of the factors of soil formation in Chapter 3 has already shown that there is a theoretical possibility of the most rapid weathering in the humid tropical regions of the world. The breakdown of the rocks and formation of the regolith has proceeded more rapidly, and has had no interruption, in contrast with the temperate regions where Pleistocene Glaciations drastically changed the climate in relatively recent times. Consequently, there is on average a greater depth of regolith in the tropical regions, with more than 30 metres recorded at some places. In the past there has been great confusion as to the definition of a soil in the tropics, confusion which has obscured present ideas of soil formation. If the soil is thought of as that part of the earth's crust influenced by current soil formation and exploited by plant roots, this does avoid the confusion between soil and parent material. However, in tropical regions as elsewhere, it is necessary to understand conditions in the parent material as these may influence processes in the soil itself.

Many inter-tropical areas have been dry land for long periods of geological history, and the deposits upon them are deeply weathered terrestrial materials dating back to the Miocene or Pliocene. Geomorphologists have recognised several surfaces associated with cycles of erosion of different ages. These surfaces have soils of different ages upon them with different profiles and properties. Some redistribution has taken place of the deposits of these surfaces resulting in a complicated pattern of soils, which can be understood only if their mode of origin is first deciphered. Concepts of erosional and depositional phases of soils, developed in Australia, have greatly assisted the elucidation of the soil pattern found on these old continental blocks (Fig. 9.1).

The climatic changes associated with the

9.1 Erosional and depositional phases of soils (after B. E. Butler)

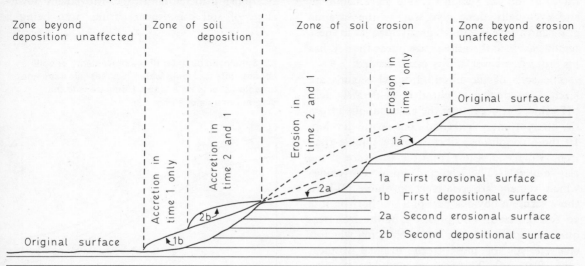

Pleistocene also affected the margins of the tropical areas. It is known from biogeographical evidence that the Sahara has been considerably moister than at present, and that the present savanna has been subjected to a drier climate. During these dry phases restricted vegetative growth led to more rapid natural soil erosion. In the moist or pluvial phases, the vegetative balance was changed with more trees growing and greater soil stability.

Humid tropical regions are characterised at the present time by a pattern of two precipitation maxima, with a total rainfall of 2000 mm. or more, and a mean annual temperature of 25 °C with only slight daily variation. This continually hot, humid climate is classed as humid tropical, continually wet (Af) by Köppen. Towards the tropics, a dry season occurs and there is only one rainy season corresponding with the overhead sun. Rainfall varies considerably with location from 600 mm. to 1500 mm. per annum, and although the average temperatures are similar to those of the humid regions, the range (20 °C) is much greater. These are called savanna climates (Aw).

The dense tropical rain forest of the humid tropical regions provides a continual supply of plant nutrients from the litter. If the cycle of nutrients from plant to soil and back to plant is broken by clearing the forest, a rapid decline in fertility is noted. Away from the continually humid regions, deciduous trees which shed their leaves in the dry season become more common, and eventually these give way to the savanna grasslands. Fire is a much greater danger in these regions as once the grasses are burnt the soil has no protective cover and is easily eroded.

The soils of the humid tropical regions are leached, producing a neutral to moderately acid pH value, but because of the bases supplied from the litter, strongly acid conditions do not develop. In these conditions, silica is more soluble than iron oxides and is lost from the structure of the clay minerals. The iron and aluminium oxides which remain are relatively insoluble. Where these form an obvious part of a soil horizon it is an *oxic* horizon. The presence of the iron sesquioxides gives the soils a red coloration which is characteristic of most freely drained tropical soils (Plate 19). The greater solubility of silica

from the clay minerals is demonstrated by the composition of drainage waters of tropical rivers. Examples are recorded where amounts of up to 50 per cent of total solids carried by rivers in the tropics consist of silica whereas the average content of all rivers only amounts to 12 per cent (Polynov). However, recent studies of weathering indicate that silica is lost in the first stage of rock weathering, while that which remains as quartz or kaolinite is relatively stable.

The character of the iron oxides remaining in the soil depends largely upon the water relationships of the soil. With freely drained soils of interfluve areas, the iron remains dispersed throughout the profile, but if ground-water is in close proximity, then the development of a strongly enriched iron horizon is more likely. However, unless the soil is desiccated, the iron does not irreversibly harden into pisolitic or slag-like concretions (pp. 72–3).

In freely drained conditions, the possibility of illuvial deposition in a B horizon is rather unlikely as the chemical environment of the deep regolith is similar throughout. In soils which have formed in a climate with a well-marked dry season, it is possible that silica may be deposited lower in the regolith. The precipitation of a silica is a fairly common feature of semi-desert areas of Australia where a silica-enriched horizon known as *duricrust* occurs where erosion reveals it at the surface.

The greater mobility of soil constituents of tropical soils is seen in the soil pattern and its relationship to the landscape. Movement downslope of soil constituents brings into being a

9.2 Schematic diagram of the movements of soil constituents including water, clay sesquioxides and organic colloids on a slope. Compare with the diagram of a catena (Fig. 5.2)

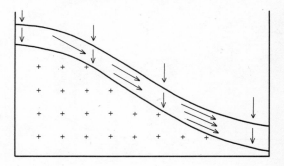

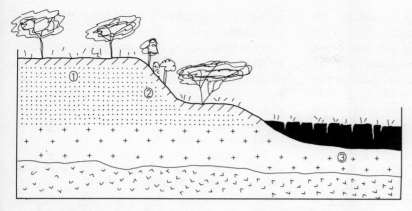

1. Reddish-brown soil developed in deeply-weathered material beneath acacia grassland. Ferrallitic soils
2. Reddish-brown soil with surface erosion developed in highly-weathered material. Ferrisols
3. Black soil of the depressions, fine-textured and salt-enriched in dry climates, deep cracking. Vertisols

9.3 Diagrammatic representation of topographic relationships of soils in the savanna lands of Africa

related sequence of soils, known as a catena (Figs. 5.2 and 9.2). Soil development in the tropical regions does not take place in a vertical direction only as is often assumed in considerations of the soils of temperate regions.

A recent soil map of Africa which has been widely accepted, distinguishes four main groups of soils characteristic of the humid tropics. These are ferrallitic soils, ferrisols, ferruginous soils and vertisols. The first three are freely drained red soils, and the Vertisols by contrast are dark in colour and poorly drained (Fig. 9.3).

Profile of a ferrallitic soil from the Congo
(Parent material – amphibolite)

Ap	0–6 cm.	Reddish-brown (2.5YR4/2) clay with well-developed crumb structure; dense root mat; sharp boundary
B_1	6–19 cm.	Brownish-red (2.5YR3/4) clay with well-developed coarse granular structure; numerous rootlets; gradual boundary
B_2	19–45 cm.	Dark reddish-brown (2.5YR3/4) clay with well-developed medium blocky structure; numerous rootlets; gradual boundary
B_3	45–74 cm.	Dark reddish-brown (2.5YR3/4) clay with well-developed medium blocky structure; few rootlets; gradual boundary
C_1	74–100 cm.	Dark reddish-brown (2.5YR3/4) clay with weak blocky structure; ferruginous concretions and weathered rock debris; gradual boundary
C_2	100–128 cm.	Similar to above, but with more ferruginous concretions and weathered debris; rare rootlets

(After Frankart)

Ferrallitic soils represent the most intensive stage of weathering and leaching as they have little or no reserve of weatherable minerals. They occur on the older land surfaces of inter-tropical Africa. These soils have little horizon differentiation and are deep, friable, porous soils with a sub-angular blocky structure. The clay minerals are of the kaolinite type with a low cation exchange capacity. These are soils with low fertility and low agricultural value. Unfortunately, 18 per cent of Africa is covered by these poor soils.

Profile of a ferrisol from Divo, Ivory Coast
(Parent material – amphibolite schist)

A_1	0–10 cm.	Dark reddish-brown humose sandy clay with fine blocky structure; roots very numerous
A_2	10–30 cm.	Dark reddish-brown sandy clay with medium blocky structure
B_1	30–110 cm.	Dark red clay with blocky structure; clay coatings on peds, fairly numerous ferruginous glossy concretions
B_2	110–200 cm.	Brighter red clay, homogeneous with few ferruginous concretions, clay coatings on peds
C	200–250 cm.	Red with diffuse greyish streaks (mottled clay) *in situ* quartz veins in the horizons between 110 and 250 cm.

(After N. Leueuf and G. Riou)

Ferrisols have a profile which resembles that of the ferrallitic soils. Structurally they are also similar, but the peds frequently have glossy surfaces caused by allumino-silica gels moving through the profile. Frequently, these soils have a structural B horizon, but they are formed where surface erosion is constantly removing the most

69

highly weathered material from the surface. As a result these soils have a slightly higher fertility than the ferrallitic soils, although still considered to be poor soils. The clays are kaolinitic, and the low cation exchange capacity is less than 80 per cent saturated.

Ferruginous tropical soils are formed where there is a pronounced dry season, for example in Africa, where they occupy 11 per cent of the surface. These soils are found on acid crystalline basement rocks beneath dry woodland and savanna in northern Nigeria. Horizon development is better than in the two preceding soils, and, as a result of leaching of iron compounds, an eluvial horizon can be seen. Sometimes a prismatic structured, textural B horizon is developed. There are often some reserves of weatherable minerals present. Although the clays are kaolinitic, the exchange capacity is higher than in both ferrallitic and ferrisol soils, particularly in the B horizon. These soils are seldom deep, and fresh rock is frequently found between 100 and 250 cm. below the surface.

Profile of a ferruginous tropical soil from Huila, Angola
(Parent material – diabase)

0–5 cm.	Very dark brown (10YR3/2) to dark grey-brown (10YR4/2) clayey sand to sandy clay, generally structureless but sometimes medium to coarse blocky structure; few ferruginous concretions; variable amounts of rootlets and medium-sized roots
5–20 cm.	Transition
20–120 cm.	Dark brown (7.5YR4/4) or brown-yellow (10YR5/4) clay or sandy clay, structureless. Sometimes a few vertical cracks. Moderately compact, almost massive, normally with a few ferruginous concretions. Variable proportion of large- and medium-sized roots, sometimes a few rootlets
120 cm.+	Parent rock

(After J. V. Botello da Costa *et al.*)

In previous classifications these soils have been called red earths, red loams or red latosols. In the recent *7th Approximation* the ferrallitic soils can be equated broadly with the Oxisols, but the absence of oxic horizons in some would necessitate their inclusion in the Ultisols. Ferrisols, if the argillic nature of the B horizon is accepted, can be placed in the Ultisols, and many of the ferruginous tropical soils would be included in the Ultustalfs.

All of these soils of the humid tropics described so far occupy positions of free drainage in the upper parts of the landscape. In lower positions, associated soils with more yellow colours occur where moister conditions prevail for longer periods. These soils have previously been called *yellow latosols* (Plate 20).

Profile of a yellow latosol from Cooroy, Queensland
(Parent material – Permian phyllites)

0–6 cm.	Brown (10YR5/3) clay loam with yellowish patches; strong crumb structure
6–15 cm.	Brownish-yellow (10YR6/5) clay loam to light clay; fine, sub-angular, blocky structure
15–30 cm.	Brownish-yellow (10YR6/6) light clay with moderate blocky structure
30–40 cm.	Brownish-yellow with few reddish mottles, light medium clay with moderate blocky structure
40–75 cm.	Mottled brownish-red (2.5YR4/6) and brownish-yellow medium clay with strong blocky structure
75–155 cm.	Mottled red, yellow and light grey medium clay
155 cm.+	Weathered phyllite

(After C.S.I.R.O.)

Profile of a vertisol from Dakar, Senegal
(Parent material – marl)

0–10 cm.	Brown-black clay, enriched with organic matter; coarse sub-angular blocky structure with glossy peds surfaces. Stable, rather porous; soil fauna very active
10–50 cm.	Black clay with well-marked platy structure; not porous; non-calcareous
50–100 cm.	Black clay, few diffuse mottles; angular blocky and platy structure, more massive, non-calcareous
100–200 cm.	Brown-black clay, small ferruginous streaks, well-developed platy structure. Numerous small calcareous nodules
200 cm.+	Marl

(After Maignien)

The lowest parts of the landscape, particularly in the tropical areas with strongly alternating wet and dry seasons, usually have dark-coloured soils which have been called black clays or vlei soils (Plate 21). The recent *7th Approximation* renamed them as Vertisols because when dry these soils crack widely and material falls down the cracks. After the passage of time, these soils *invert* themselves by this process. Similar black soils which develop strong micro-relief features have been described from Australia, where the phenomenon

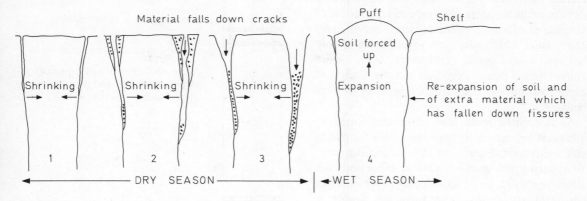

Material falls down cracks   Puff   Shelf

Soil forced up

Shrinking   Shrinking   Shrinking   Expansion   Re-expansion of soil and of extra material which has fallen down fissures

1   2   3   4

——————— DRY SEASON ———————|←— WET SEASON —→

9.4 Gilgai phenomena in Vertisols. Shrinkage during the dry season allows material to fall down the fissures. Expansion when wetted in the wet season results in the soil being forced up to form a ridge (puff) whilst the area around the fissures remains at a lower level (shelf)

was called *gilgai* (Fig. 9.4 and Plate 22). The strong cracking is a property of the montmorillonitic and mixed layer clays which have a high cation exchange capacity. The better drained of these soils, developed upon basic rocks rich in ferromagnesian minerals, are amongst the best in Africa for agricultural use.

Many of these black soils remain undeveloped and are used only as low-quality grazing land. If cultivated and where necessary irrigated, cotton, sugar-cane, as well as corn, wheat, sorghum and rice could be grown. Where drainage is poor and the climate drier, the accumulation of salt can be a problem and saline or alkaline soils may occur. In North Africa Vertisols associated with depressions in the limestone areas have already been noted. However, it is within the inter-tropical areas that these soils attain their widest distribution.

The name *krasnozem* is given to tropical and sub-tropical red soils developed upon base-rich parent materials. Krasnozems are named from the southern part of the U.S.S.R. where they occur in the Trans-Caucasian part of Russia. These soils have been identified also in the Hawaiian Islands and in Australia where they are developed upon basalt parent materials in Queensland and northern New South Wales (Plate 23). Originally developed under tropical rainforest, such freely drained soils lack horizon development because of the flocculating effect of their high content of hydrated ferric oxide.

Profile of a krasnozem from Springbrook, Queensland
(Parent material – basalt)

0–25 cm.	Dark reddish-brown (5YR3/3) light clay with granular to fine blocky structure
25–135 cm.	Red (2.5YR3/6) medium to heavy clay with blocky structure
135–220 cm.	Brownish-red passing into reddish-brown (5YR4/4) heavy clay with blocky structure
220–250 cm.	Reddish-brown with brownish-yellow and light grey heavy clay with blocky structure
250–275 cm.	Yellowish-brown, light red, etc. clay loam with lumps of brittle, weathered basalt
	(After C.S.I.R.O.)

Krasnozems are fertile soils used for sugar-cane and pineapple in Queensland as well as for other fruit and vegetable crops.

Profile of a 'giant podzol', Mackenzie, Guyana
(Parent material – Berbice formation, white sand)

0–40 cm.	Spoil
40–55 cm.	Very dark grey (5YR3/1) sandy clay
55–70 cm.	Very dark grey (5YR3/1) sand
70–125 cm.	Bleached white sand
125–140 cm.	Transitional to humic horizon
140–165 cm.	Black humic sand
165–180 cm.	Transitional loamy sand, very hard, moist
180–200 cm.	Gleyed sandy loam resting upon iron pan 1 cm. thick
200–231 cm.	Light red (2.5YR6/5) sandy loam
	(After Bleackley and Khan)

Podzolization can be seen in the tropics. In Malaya, Guyana and the Congo deep 'giant' podzols have developed in alluvial sands on river terrace sites where free drainage enables profile development to occur. An impoverished form of tropical rainforest, the 'heath forest' occurs upon these soils which are very much more deeply leached than similar temperate soils. More conventional podzol profiles can be found at higher altitudes in the mountain areas of tropical lands.

Soils formed from both river and estuarine alluvium are extremely important soils for paddy rice cultivation. Because of the continued saturation during the growing season, these soils are strongly gleyed, and iron sulphides may occur as concretionary forms. The following profile is typical of a mangrove swamp soil.

Profile of a soil on mangrove swamp, Belo, Madagascar
(Parent material – fluvio-marine alluvium)

0–60 cm.	Yellowish-brown, clayey, plastic and adherent, rich in roots and rhizophores
60–100 cm.	Progressive transition to a greyish-blue horizon with sulphur yellow or rusty-orange mottles and cavities more or less hardened (iron sulphides), fine sandy clay, plastic
100 cm.+	Pale grey, fine sandy, rich in mica and dark minerals

(After Hervieu)

High temperatures, with rapid rates of organic decomposition, limit the formation of organic soils in tropical lowland areas. However, organic soils are reported from the lowlands of Sumatra and Borneo. The peat has formed from the debris of trees, unlike the moss peat of temperate climates.

So far, this discussion of the soils of the humid tropics has avoided as far as possible the use of the term laterite. *Laterite* is a phenomenon widespread in tropical areas, and is formed by an accumulation in the soil of sesquioxidic material. In most cases the appearance of laterite crusts is a relic feature of a previous episode of soil formation. A *lateritic horizon* is formed where the ground-water movements within the soil concentrate iron and aluminium oxides into a restricted layer (Fig. 9.5). These concentrations may appear as nodules (Plate 24), as a cellular mass or as a slag-like accumulation (Plate 25). In all cases where an indurated crust occurs, it seems to have suffered drying and irreversible hardening. The

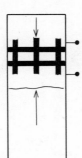

Leaching down to water-table

Concentration of iron and alluminium in nodules or slag-like concretions

Water-table steadily rising causing migration of iron and aluminium into zone alternately wet and dry

Erosion removes soil down to concretionary zone which irreversibly hardens to form crust

9.5 Formation of laterite

correct usage of the term laterite is for 'a massive vesicular or concretionary ironstone formation nearly always associated with uplifted peneplains originally associated with areas of low relief and high ground-water'. Uplift of the land surface has resulted in increased fluvial erosion by streams which have cut deeper valleys and lowered the water-table. Where it has been revealed by erosion the ironstone horizon has become irreversibly hardened by contact with the air (Plate 26). Erosion has also stripped off the overlying leached horizons so that the laterite usually occurs as a plateau remnant in an interfluve position. Good examples of this occur in Western Australia where the relic soils form part of an eroded Pliocene or Miocene surface (Plate 27). The breaking up of the laterite has formed new parent materials for soils in the present phase of soil formation. Even on the crust itself, some weathering has taken place to form a new thin soil in which a poor scrub vegetation grows. Below the crust are the mottled and pallid zones. The mottled zone is subject to some alternation of conditions, whereas the pallid zone has been completely saturated and contains only kaolin clay, the iron having been moved in a reduced state into the zone of the present indurated crust.

Laterite remains soft while moist and in a fresh state can be cut into blocks and then dried in the sun. Once dried and hardened it can be used as building material, for example the Angkor Wat temples in Cambodia. If a soil is stripped of its forest cover and exposed to the strong rays of the sun, it is argued that an irreversible hardening takes place, and that the areas of laterite are being extended by agricultural activities in the tropics. Where there are thin crusts, the crust can be broken and planted with trees, but whatever crop is considered, the nutrient supply of these soils is low.

Profile of a laterite from near York, Western Australia
(Parent material – gneiss)

0–150 cm.	ironstone crust	Yellowish-brown hard ironstone, becoming slightly softer towards the base
150–690 cm.	pallid zone	White and slightly pinkish clay with dark red patches. Quartz grains are bleached white in the clay and stained with iron oxide in the red. Occasional mica flakes throughout
690 cm. +	transition to parent rock	Pale brown and rusty mottled, weathered gneiss

(After M. J. Mulcahy)

10 INTRAZONAL AND AZONAL SOILS

Although the zonal soils considered in the previous chapters have their development influenced by all the factors of soil formation, that of climate exerts an overriding influence producing a pattern which is roughly comparable with the climatic zones. Within the various zones are areas of well-developed soils, which reflect, however, the local dominance of a single factor such as parent material or drainage conditions. When a particular type of parent material exerts a strong influence over soil formation, as in the case of limestones, soils with *calcimorphic* characteristics are developed. In a similar manner, the continued presence of water in the soil causes the development of the features of gleying associated with *hydromorphic* soils. The presence of soluble salts in the soil confers upon it chemical, physical and biological features which require a special consideration in any classification. These are the *halomorphic* soils. These three groups of soils, calcimorphic, hydromorphic and halomorphic, form the main divisions of the intrazonal order of soils. Already they have been mentioned in the consideration of the associated zonal soils, but for completeness they are considered together in this chapter.

Calcimorphic soils

The stability conferred upon a soil as the result of the presence of calcium has already been commented upon (p. 13). This property, together with the slightly alkaline pH values, results in a particular profile form and justifies the classification of calcimorphic soils as intrazonal. The profile of a soil on a limestone, compared with an adjacent soil on a non-calcareous parent material, is usually less leached and lacks strong horizon differentiation.

Several features of calcareous soils make them distinct from brown earths with which they can be linked in a maturity sequence. In the first place, the vegetation growing upon these soils is usually a form which produces a leaf-litter which is rich in bases. Thus, there is a continual return of bases to the surface of the soil. Secondly, the faunal population of these soils is numerous, encouraged by the more nutritious leaf-litter. Thirdly, the presence of calcium-saturated clays and free calcium carbonate in the soil inhibits the movement of soil constituents by the formation of stable calcium compounds which remain flocculated. Fourthly, because of their stability these soils are usually fairly rich in organic matter, have black or dark reddish-brown colours and stable structural aggregates in the form of crumb or blocky peds. Lastly, because they are formed upon rocks which have little insoluble residue, these soils are usually shallow and have low moisture reserves. As they occur over limestone, calcareous soils are invariably freely drained; however, on calcareous boulder clays transitional soils to hydromorphic soils occur, called calcareous gley soils. The two main types of profile seen are the *rendzina* and the *brown calcareous soil* (Fig. 7.9). The name rendzina, which is derived from a Polish peasant name, is widely used. Soils of this nature are described from almost all parts of the world from the temperate regions to the humid tropics. Brown calcareous soils are described from Britain and Europe, where there is some overlap with the name brown forest soils.

The rendzina is a shallow soil rich in organic matter and biological activity. It has a stable crumb structure and is dark in colour. It is a relatively simple soil with an A horizon directly overlying a C horizon which is the limestone parent material. The humus is well incorporated in the mull form, and micro-morphological evidence shows that these soils are largely composed of the faecal pellets of soil arthropods and the casts of earthworms (Plate 28).

Profile of a rendzina from the Hartz foothills, Germany
(Parent material – chalk)

A 0–25 cm. Brownish-black, strongly humose, calcareous stony clay loam with crumb structure
C 25 cm.+ Greyish-white, laminated fissured chalk

(After Muckenhausen)

Profile of a brown calcareous soil from Nottinghamshire, England
(Parent material – Permian limestone)

Ap 0–23 cm. Dark reddish-brown (5YR3/2) sandy loam with occasional angular fragments of limestone and rounded quartzite stones; fine to medium sub-angular blocky structure; friable; moderate amount intimate organic matter; abundant fibrous roots; sharp boundary
(B) 23–35 cm. Reddish-brown (5YR4/2) sandy clay loam with occasional fragments of limestone and rounded quartzite stones; weak medium sub-angular blocky structure; friable organic matter confined to earthworm channels; calcareous; sharp boundary
C 35 cm.+ Weathering limestone

(After E. M. Bridges)

Brown calcareous soils are characteristically formed over Jurassic limestones in Britain, but they are also described from Africa, U.S.A. and other areas of the world. These soils develop on limestones with a larger insoluble residue, and therefore they become deeper than the rendzina having a (B) horizon developed between the A horizon with its mull humus and the limestone C horizon. The A horizon is usually dark reddish-brown, crumb structured and moderately rich in organic matter, with a neutral or slightly acid reaction. The (B) horizon is distinguished by a more ochreous coloration caused by stable iron compounds. Although these soils are deeper than the rendzinas, they are seldom more than 75 cm. from the surface to the rock beneath. This gives them greater moisture reserves, and they are considered to be good agricultural soils.

Because of their shallowness, both rendzinas and brown calcareous soils may not have supported dense vegetation in the past. In Britain, the rendzina is typical of the downland of southeast England, but examples can be seen also where erosion has reduced the depth of soils on other limestones. Areas of brown calcareous soils in Britain were probably scrub woodland and many limestone districts still bear the name 'heath' as in the county of Lincolnshire. It is only in relatively recent times that these soils have proved to be excellent arable soils.

Hydromorphic soils

Poor drainage can be observed in the soils of most regions of the world and as such represents the most widely spread of the processes of soil formation, leading to the formation of *gley* or *hydromorphic soils*. Often these soils are analogous with soils of similar parent material on freely drained positions of the landscape, and a complete range from freely-drained to poorly-drained soils can be seen (Fig. 5.3). Hydromorphic soils can be found in association with all zonal soils, anywhere in fact where water can gather in sufficient volume and for sufficient time to produce the effects of gleying.

Gleying occurs when water saturates a soil, filling all the pore spaces and driving out the air. Any remaining oxygen is soon used by the micro-biological population, and anaerobic conditions are established. At the same time, the soil water contains the decomposition products of organic matter. In the reducing conditions brought about by the absence of oxygen and in the presence of organic matter, iron compounds are chemically reduced from the ferric to the ferrous state. In the ferrous form iron is very much more soluble, and is removed from the soil leaving the colourless minerals behind. This gives gley soils their characteristic grey coloration.

Hydromorphic soils can be sub-divided into those which have continually saturated conditions, and those which have a temporary period of saturation only. Generally, this division distinguishes those soils with a permanent water-table within the soil from those that are slowly permeable. European pedologists have referred to the former as *gley soils*, and the latter as *pseudogley soils*. In Britain these have been called *ground-water gley* and *surface-water gley* soils.

Surface-water gley soils are those in which the drainage is impeded above an impervious or very slowly permeable sub-soil horizon (Fig. 10.1 and Plate 29). This leads to the development of grey colours along the fissures and pores of the soil, particularly in the B horizon and the lower parts of the E horizon. Usually these soils are developed from fine-grained parent materials in which clay movement has taken place to form a textural B horizon. However, not all surface-water gley soils have clay skins in their B horizons, and the calcareous gley soils in particular differ from them in this way.

Profile of a surface-water gley soil from Derbyshire, England

(Parent material – glacial drift)

L		Discontinuous litter of beech, sycamore and oak leaves
F	2·5–0·5 cm.	Comminuted leaf fragments, darker and more humified towards the base of the horizon
H	0·5–0·0 cm.	Black amorphous humus
A	0·0–1·5 cm.	Very dark grey (10YR3/1) stoneless sandy loam with bleached sand grains; medium sub-angular blocky structure; friable; high amount organic matter; abundant fibrous roots; narrow irregular boundary
Eb	1·5–23 cm.	Brown (10YR5/3) slightly stony sandy clay loam with medium sub-angular blocky structure; common woody roots; narrow boundary
Ebg	23–35 cm.	Mottled yellowish-brown (10YR5/4) to strong brown (7.5YR5/8), with greyish-brown (10YR5/2) on structure faces; slightly stony sandy clay loam with medium angular blocky structure; firm; low amount organic matter, few roots; earthy iron and manganese concretions; narrow boundary
Bg	35–75 cm.	Dark brown (7.4YR4/4) mottled to pale olive (5Y6/3) on structure faces; slightly stony clay with coarse prismatic structure; firm, plastic when wet; low amount organic matter, few roots; black manganiferous patterning; merging boundary
Cg	75 cm.+	Dark brown (7.5YR4/4) mottled to pale olive (5Y6/3) stony firm clay with structure no longer obvious; firm, plastic when wet; no visible organic matter, no roots; slightly calcareous

(After Bridges)

Most gley soils may become aerated occasionally in the event of a prolonged drought. As a result they may have the colours of ferric iron

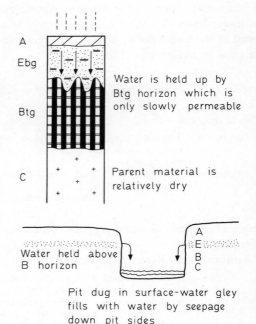

10.1 Diagrammatic profile of a surface-water gley soil

compounds present as a mottling in the Bg horizons. Gleying can best be seen along the fissures and pores where the effects are most concentrated; the internal parts of the peds often remain aerated and in the ferric state. In the A horizon of most gley soils, grass roots frequently become coated with iron in the form of rusty sheaths.

Ground-water gley soils include those in which there is a water-table which rises to within 60 cm. of the soil surface (Fig. 10.2). Usually these soils are formed from rather permeable parent materials such as alluvial sands and gravels which overlie an impervious sub-stratum upon which water accumulates. Therefore, these soils occur in the lower parts of the landscape and are often transitional to organic soils. In the natural state these soils have well-developed grey colours caused by continual anaerobic conditions, but as many of these have been artificially drained mottling can often be seen.

Halomorphic soils

In semi-arid and arid parts of the world soils are developed under the influence of soluble salts or with sodium as the dominant exchangeable cation. These are classified as halomorphic soils. The presence of soluble salts or exchangeable sodium exerts an adverse effect upon the growth of most crops, producing specific physical features in the soils. As salty soils usually occur in lowland alluvial sites, they coincide with potentially fertile and adequately watered areas.

Halomorphic soils occur in all continents as discontinuous patches within other major soil groups. As they do not form a zonal type of their own, they are classified with the intrazonal soils (Plate 30). Two conditions are necessary for salt accumulation in soils; firstly, a dry climate in which salts are not leached out and secondly, a parent material or ground-water which contains salt. The salt can be derived from rock salt or from the weathering of sodium silicates. Alternatively, salt can be obtained from salt spray which is driven inland into an arid area and not leached away. Soils can undergo a secondary enrichment of salt by artificially raising the water-table so that water can evaporate from the soil surface, leaving salts behind. Examples of the secondary enrichment of soils have occurred in the Indus Valley, Pakistan and in California, U.S.A. In these and other areas affected by salts, either crops are reduced in yield or the land is eventually abandoned as being useless. Remedial action is often practicable, but in many cases is not financially possible.

Profile of a solonchak soil from Kayseri, Turkey (Parent material – lacustrine alluvium)

0·5–0·0 cm. White (10YR8/2), dry salt crust
0·0–20 cm. Pale brown (8YR6/3) very weak granular or structureless silty clay; friable; hard when dry; strongly calcareous; gradual boundary
20–55 cm. Same colour and texture; firm, hard, rounded vesicular aggregates about 1·5 cm. diameter; strongly calcareous with numerous salt crystals and eyes or spots of salts
55–105 cm. White (5Y8/2) silty clay; saturated below 80 cm. – apparently level of water-table at time of sampling.

(After Oakes)

The salts which affect soils are chiefly the sulphates, chlorides and carbonates of sodium and magnesium. Salts can be present either as an electrolyte in the soil solution or as cations which occupy exchange positions on the clay-humus

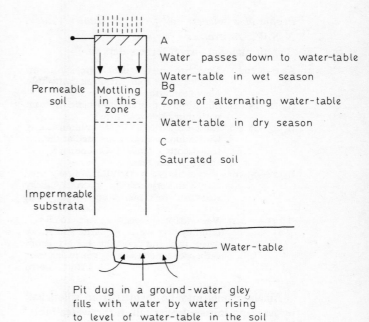

A
Water passes down to water-table

Water-table in wet season
Bg
Zone of alternating water-table

Water-table in dry season

C
Saturated soil

Pit dug in a ground-water gley fills with water by water rising to level of water-table in the soil

10.2 Diagrammatic profile of a ground-water gley soil

complex. Soils in which there are concentrations of neutral salts, such as sodium sulphate or sodium chloride, have a lower pH value, generally less than pH 8. Although sodium is present, dispersion does not take place. The soils are reasonably well-structured. They may have an efflorescence of salt on the surface during the dry season, a feature which gives them their American name – *white alkali soils*. This type of soil is equally well known as *solonchak*, the Russian name for these soils (Fig. 10.3).

10.3 Solonchak soils. Salts are carried into the soil by salt enriched ground-water. The water evaporates from the surface leaving a salt crust

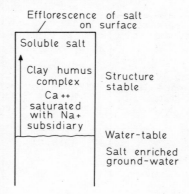

Efflorescence of salt on surface

Soluble salt

Clay humus complex
Ca ++ saturated with Na+ subsidiary

Structure stable

Water-table

Salt enriched ground-water

Profile of a solonetz soil from the Hunter Valley, N.S.W., Australia

(Parent material – river terrace deposits)

A 0–8 cm.   Dark to very dark greyish-brown (10YR4/2) friable silt loam with thin platy structure

A₂ 8–15 cm.   Pinkish-grey (7.5YR7/3) friable silt loam with thin platy structure

B₁ 15–28 cm.   Dark brown (7.5YR3/2) columnar clay with columns weakly rounded at the top and slightly degraded. Columns are 6 to 10 cm. in diameter

B₂ 28–75 cm.   Very dark brown (10YR2/2) heavy hard clay with very coarsely irregular blocky structure, very plastic and sticky when wet

B₃ 75+ cm.   Very dark greyish-brown (10YR3/2) crumbly clay which is plastic and sticky when wet and contains hard calcium carbonate concretions 0.5 cm. in diameter

                                  (After Thorp)

Alkaline soils may be formed from solonchak by leaching initiated by a fall in the water-table, increased rainfall or irrigation. As the leaching of sodium ions from the soil includes some special reactions, the process has been given the name of *solodization*. The soluble salts are easily washed out, and if the soil is dominated by calcium ions, it reverts to the soil appropriate to its region – a chernozem or chestnut soil. If dominated by sodium ions, there is a loss of stability so that the surface horizons become structureless with dispersed humus and clay. This gives these soils a dark-coloured appearance which caused them to be called *black alkali soils* (Fig 10.4). Alternatively known by their Russian name of *solonetz*, these soils have a high pH value brought about by the presence of sodium carbonate in the soil according to the following reaction:

$$Na \text{ clay} + H_2O \rightarrow H \text{ clay} + NaOH$$
$$2NaOH + CO_2 \rightarrow Na_2CO_3 + H_2O$$

Hydrogen ions from rainwater displace sodium ions from the exchange positions forming acid clays, and sodium hydroxide appears in the soil solution. Carbon dioxide is present in the soil, released from plant roots and bacteria; it combines with the sodium hydroxide to form sodium carbonate which raises the pH value to pH 9 or greater. The dispersed clay and humus move down the profile which leads to the development of a surface horizon which is dark grey. Below this is a light grey platy or structureless eluvial horizon which overlies an extremely intractable clay B horizon arranged in the columnar structures so typical of these soils (Plate 31).

There is a complete range of soils from the solonetz to a soil which is completely free of sodium ions. The intermediate soils are known as *solodized solonetz* and the completely leached soil is known as a *solod*. The A horizons of the solodized solonetz are deeper and they possess caps of amorphous silica on top of the columnar structures of the B horizon. This solodization process is evident from the description of the solonetz soil given above, when the columns are described as being slightly degraded. When the complete removal of sodium ions is achieved, the soil becomes acid, the structure stabilises, and cultivation is again possible (Fig. 10.5). Unfor-

10.4 Solonetz soils. Lowering of the ground-water enables loss of calcium, this results in saturation by sodium alone and instability of structure

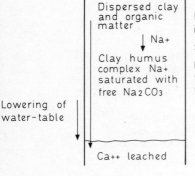

10.5 Solod soil. Complete leaching results in a solod soil which is acid and moderately well structured. 1. Leaching of organic matter. 2. Leaching of calcium ions. 3. Leaching of soluble salts

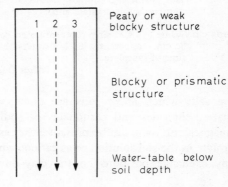

tunately the amelioration of these soils is patchy in occurrence and a uniform reclamation is difficult to obtain as halomorphic features persist in the lower-lying parts of the landscape.

Profile of a solod soil from Askania Nova, Ukraine (Parent material – redistributed loess-like loams)

A$_1$	0–10 cm.	Grey, just moist medium loam with platy structure; friable and non-calcareous; gradual boundary
A$_2$	10–30 cm.	Pallid grey medium loam with platy structure; friable and non-calcareous; sharp boundary
A$_2$Bg	30–100 cm.	Greyish-olive-green medium loam with nutty-prismatic structure compact and non-calcareous; iron-manganese concretions; gradual boundary
Cg	100–230 cm.	Olive-green clay; very compact and plastic; abundant iron-manganese concretions. From 130 cm. compact calcareous concretions

(After Grin and Kissel)

In several parts of semi-arid Australia salting has occurred on the lower slopes and valley bottoms as the result of the interference in the landscape by man. Removal of the original forest resulted in an increased rate of removal of salts from the soils of the interfluve areas, but the rainfall is not sufficient to flush the salt right away. Consequently, it accumulates in the valley bottom lands rendering them unsuitable for most plants.

Knowledge of halomorphic soils has been of use in areas outside the desert and semi-desert regions of the world. Reclamation of the soils of the polders of Holland has necessitated treatment for the effects of salts.

Other soils which are worthy of mention include the *Andosols*, which are named from soils developed on recent base-rich volcanic deposits in Japan. These soils are also intrazonal in that they occur in many different zones from Alaska in the tundra to tropical Africa and Indonesia. Their colour is dark brown and they have a thick friable, organic-rich A horizon overlying a (B) horizon which shows little development of clay movement. The Andosols have a low bulk density and they are not sticky. *Brown soils* are also described from Africa which occur on recent volcanic ash, basic rocks or alluvial deposits. As these are relatively young soils, they contain a

reserve of weatherable minerals, and their clays are montmorillonitic. Although not extensive in area, these are fertile soils used for crops of bananas, cocoa and coffee, though their position on steep slopes often limits their usefulness.

Azonal soils

Azonal soils may be found within any of the zonal soil distributions and can be distinguished from them and from the intrazonal soils in that they lack well-developed soil characteristics. Some factor such as youthfulness, the parent material or relief has prevented the development of well-defined pedological features. In all cases these soils lack a B horizon, and the thin rudimentary A horizon is only distinguished from the C horizon by the presence of organic matter. Immature soils have caused problems in previous soil classifications which dealt with mature soils only. However, recent attempts have established positions for azonal soils in a world classification. In the American *7th Approximation* these soils are included in the order Entisols, but their definition includes lithosols, regosols, alluvial soils as well as thin rankers.

Lithosols develop at high altitudes where resistant parent materials withstand the disruptive forces of weathering. Soil formation is slight, resulting in a humus-enriched shallow soil which is stony with very little fine earth. The lithosols are found in exposed sites where natural erosion is active, such as in mountain areas. All early stages of soil formation leading to *rendzinas* on calcareous parent materials and to *rankers* on non-calcareous parent materials, can be considered as lithosols.

Regosols are developed upon deeper, unconsolidated parent materials such as dune sands or volcanic ash. Elementary profile development takes place rapidly through the highly permeable sand, and although superficial organic horizons may form as well as an A horizon, these soils lack any illuvial horizons. Because of the mobility of the parent material, multiple profiles may be seen where fresh material has accumulated over an already existing regosol profile.

Alluvial soils are variable in texture, drainage and state of maturity. They are liable to flooding and the surface receives fresh additions of material which are laid down in successive layers, often of different grain sizes. Some alluvial soils are poorly drained, even peaty, but others on levees and terraces are imperfectly or freely drained. As they are water-deposited, many alluvial soils retain their layer nature, but older terrace soils gradually achieve maturity and come to resemble adjacent upland soils.

Alluvial soils in many parts of the world have been altered and cultivated by man from very early times. In tropical regions, paddy fields have been constructed from river alluviums, in desert regions irrigation has been practised, and in temperate regions the level of alluvial land has been raised by encouraging artificial sedimentation, a process known as *warping*. Marine alluviums can be reclaimed from the sea, but with these there is the added problem of leaching out the salt from the newly embarked sea-marsh. It has been found necessary to mix different layers of sediment to form a reasonable soil texture in some of the Dutch polders.

Mountain soils

The soils of mountainous regions are frequently shallow and subject to erosion on steep slopes. Different aspects can have a considerable influence upon the soil profile developed on either side of a valley. The increased altitude brings different climate and vegetation formations, so it can be expected that soils will also occur in altitudinal zones in mountainous areas. Several writers have described sequences from different parts of the world including the Big Horn Mountains of America which have a sequence of grey desert soils, brown desert soils, chestnut soils, chernozems, prairie soils and podzolic soils. In the tropics, a sequence from ferrallitic soils on the shore of Lake Tanganyika to the snow-line is given in Fig. 10.6. A similar sequence is given for the Tien Shan Mountains, which passes from the chernozems to the snow-line (Fig. 10.7).

Organic soils

Hydromorphic and halomorphic soils are both developed on poorly drained sites, often in association with organic soils (Histosols) on very poorly drained sites. An arbitrary boundary between true peats and the peaty mineral soils is usually drawn where the peat depth exceeds 38 cm. in Britain, but elsewhere other depths are used. Knowledge about organic soils is not plentiful, but the evidence available suggests the division into acid and alkaline or neutral peats. Peats can also be classified according to the site in which they are developed.

Accumulations of organic matter are encouraged by wet conditions; these result from heavy rainfall, seepage or flooding, and high levels of ground-water. Where rainfall is generally in excess of 1500 mm. per annum in the uplands of Britain, *blanket bog* peats are favoured, while *raised bog* occurs in wet sites on the lowlands. As both of these forms are above the water-level, they rely entirely upon rainwater for their supply of moisture. Seepage waters and high ground-

10.6 Mountain soils. Vertical zonality of soils on Kivu, East Africa

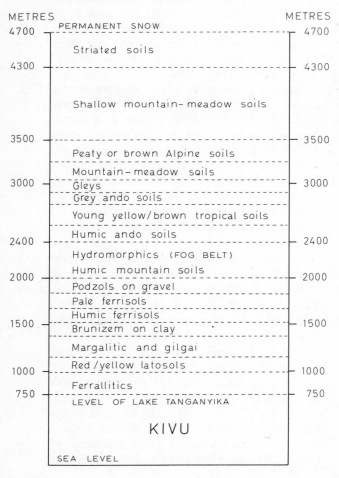

waters are responsible for the accumulation of lowland peat forms. Different water-levels in the past have resulted in different types of peat developing in any specific area. These may be peats formed from a reed-swamp community of *Phragmites*, or carr peat which has included within it woody fragments of birch, alder and willow which were able to grow in slightly drier conditions. Raised bog and blanket bog peats frequently have formed from a *Sphagnum-Eriophorum* community of plants (Fig. 10.8).

The most important form of lowland peat is *fen peat* which is developed where base-rich waters are associated with an accumulation of organic matter. Peat with a neutral or slightly acid reaction gradually develops as in the fens of England. Fen peat is characteristically black or very dark brown and there are few recognisable plant remains preserved in it.

Acid lowland peats can be formed in places

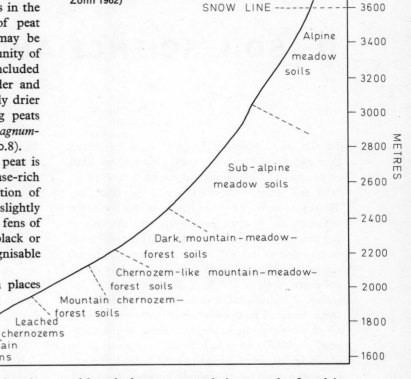

10.7 Mountain soils. An altitudinal sequence on the N.E. slope of the central Tien Shan (after Zonn 1962)

with drainage from non-calcareous rocks; in Britain examples can be found in Lancashire and Somerset. Where rainfall is sufficient raised bog can develop overlying both fen and acid lowland peats. Raised bog peat is formed of brown peat with little mineral matter, and a high proportion of recognisable plant remains. It is built up in the form of a low dome and is usually strongly acid in reaction.

On the uplands of western Europe, particularly western France and Britain, blanket bog covers large areas over 300 m. above sea level. Blanket bog often attains a depth of about 150 cm,

although deeper accumulations can be found in declivities. It is black, strongly acid and consists of few identifiable plant remains with very little mineral matter included (Plate 32). It supports a growth of cotton grass and sphagnum moss, but with drainage ditches heather may become dominant and the area can then be grazed.

While the upland peat moors are of little agricultural value because of their extreme wetness and acidity, lowland peat areas are valuable and versatile agricultural and horticultural land. Liming can cure the acidity but this may bring trace element deficiencies into evidence.

10.8 Basin and raised peat formation

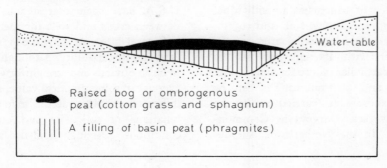

81

11 SOIL SCIENCE AND LAND USE

In the past, knowledge about soils has been acquired slowly, and only communicated by word of mouth. The use of the soil has been evolved by trial and error controlled largely by tradition. However, in recent years the necessity for more planning has become paramount. By the end of the twentieth century the maximum use will have to be made of our soil resources, particularly for the production of food. As yet we have only vague information about the soils of many parts of the world. Although most countries now undertake some research in pedology and have some form of soil survey, these organisations are usually small bodies with much to do. New developments in the use of sensory methods of recording, using the reaction of organic content, heat and moisture content, instead of straightforward photography, offer the possibility of more rapid mapping techniques in the immediate future. In spite of this it may be two or three hundred years before most countries are mapped in detail; a lamentable situation at a time when demands for food are far outstripping the supply.

In those countries with large areas still undeveloped or at a low level of production, the apportionment or reapportionment of the land could be an important sequel to natural resource surveys, including soil surveys. Perhaps the best known series of surveys of this type are those done by the Land Research Division of C.S.I.R.O. in Australasia. The landscape is analysed into 'land systems' each of which has an individual pattern of geology, topography, soils and vegetation. In this way an undeveloped area can be assessed, together with its climate, for its potentialities. In particular its use for agriculture, grazing, forestry and communications can be forecast. Similar surveys are carried out by the Directorate of Overseas Surveys in Commonwealth Countries. In the Netherlands, surveys have been made of the newly drained polders of the Ijsselmeer, where land-use plans have been drawn up for the lands originally beneath the sea.

Arid and semi-arid countries have limited food production because of a lack of water and facilities for irrigation. Although in many cases water can be made available, it is necessary first of all to assess the soils of the area for their suitability for irrigation and the soluble salt content of the water supply. The topographic situation of the scheme must be considered, for water must be brought easily to the site. The under-drainage must be satisfactory or water-levels will rise and salinity may well occur.

When dealing with soils the agricultural aspect always comes to mind first. A soil survey gives an adequate knowledge of the distribution of the soils and their chemical and physical properties. This enables more accurate advice to be given to the farmer regarding fertiliser application to amend plant nutritional deficiencies and lime to correct acidity. At the present time little information is available in Britain about the yield of crops in relation to the specific soils on which they are grown. Figures for overall production can be obtained and farmers are well aware of the productivity of their own land. However, any field or group of fields usually includes several different soils with varied properties and fertility. Some more detailed investigations have been made recently in Britain, but soil scientists in the U.S.A. have been interested in the relationship between crops and soils for many years.

Several attempts have been made to classify land and its capability. Although this might seem to be an obvious and straightforward geographical exercise of some practical value, it is fraught with difficulties because of the many different combinations of soil, site and current economic conditions. Although land-use maps have been used in Britain for planning purposes, they are not based upon the fundamental properties of the land. However, if a land classification is based upon the fundamental properties or limitations imposed by the environment, which are not so easily changed, the resulting classification can be interpreted in the light of current economic conditions. Limitations included in this type of soil and landscape assessment include: wetness caused by impermeable or slowly permeable soil horizons, high rainfall or flooding; shallowness, stoniness, extremes of texture and structure and inherent low fertility of the soils; gradient of the land and the soil pattern; liability for erosion; climatic limitations induced by increasing altitude above sea-level and increasing rainfall. Consideration of these limiting factors enables land to be classified on a scale of eight classes (U.S.A.) or seven classes (Great Britain). In both schemes arable land with diminishing versatility comprises classes I to IV. The remaining classes are concerned with land most suited for grassland, forest and non-agricultural natural reserves. Soils in these classes are expected to respond to management and improvement in similar ways.

Increasing knowledge of the soil may in future take a larger part in the valuation of land for agricultural and horticultural purposes. Obviously the soil already plays an important part in a prospective buyer's assessment of a market price, although his assessment is usually based upon instinct rather than upon an accurate knowledge of the soils on a property. Perhaps in future years the Inland Revenue may use the land capability category as a basis on which to assess tax liability.

As the natural timber resources of the world are constantly being consumed many countries have to consider replanting their forests on a fairly large scale. Whilst many foresters know by experience where certain species of tree will grow satisfactorily, newly acquired areas can be planted with a much greater confidence of success if conditions of the soil site and semi-natural vegetation are known in advance.

In times of strategic necessity, information can be provided by soil maps about the feasibility of the terrain for military vehicles, or for the location of roads or airfields. In times of peace, the same interpretation can be used for civil engineering undertakings.

The soil scientist can assist in the reclamation of eroded and derelict land. Large areas of the world are affected by salinity or have suffered soil erosion through over-exploitation, as in America, Russia and Brazil. The maintenance of the physical characteristics of the soil are as important as its fertility, as with a breakdown of structure soils are liable to be eroded away easily. Soil scientists have co-operated with farmers and agricultural advisors to work out the best cropping and cultivation programme commensurate with soil conservation. Terracing, strip-cropping, shelterbelts, special types and times of cultivation and maintenance of reasonable levels of organic matter are all encouraged by soil scientists in an attempt to counteract soil erosion by wind and water. Areas which have been mined by the opencast method for coal, lignite, ironstone, gravel and many other minerals can be restored to a productive capacity. Soil scientists in collaboration with agriculturists, foresters and ecologists can help to accelerate the natural processes of revegetation and return to usefulness.

Town and country planners could use information on the soil at all levels of decision-making. At the level of regional planning, the overall policy will require information to enable land to be allocated for residential, commercial, industrial, agricultural and recreational development. At a local level the siting of houses, schools and other buildings may well be determined by the soil distribution, particularly if cracking, heaving clay soils occur. In rural areas, away from a mains sewerage scheme, disposal of sewage in septic tanks can provide problems if the soils and subsoils are not permeable. In the past, planning of suburban areas has taken too little cognisance of the natural environment in the development of new housing areas. The American soil scientist Kellogg has written: 'People have no need whatever to put their houses where they will slide down hill, settle and crack, or be flooded; nor where their gardens will be contaminated with sewage effluent; nor where their homes cannot be beautified with growing plants. It has been demonstrated that what has been learned about soil selection for the many specialities in farming can be used for these other purposes.'

A greater knowledge of the soils of the world brings with it a better insight into ecological

83

relationships. The dependence of agriculture and forestry on the soil has already been mentioned, but equally important are the nature reserves and water catchment areas which are of great amenity value. The soil scientist can contribute a valuable opinion on the mangement and use of such areas. Frequently, these areas serve a dual purpose. With increasing leisure time, the management of recreational land becomes more important. Soil surveys can give information as to the best areas for camp sites, games areas and trails throughout nature reserves.

Pedology is a correlative science, bringing together the many facets of the environment as they are involved in the formation and maintenance of the soil. It is a young science and has within it room for many different scientific approaches. The information of soil science can be used by many people in agriculture, archaeology, civil engineering, plant and animal ecology, and forestry. It should play a greater part in geographical appraisal and the planning decisions of local and national authorities.

BOOKS AND JOURNALS FOR FURTHER READING

There are numerous textbooks on the subject of pedology, some of which are mentioned below. Information regarding soils is also published by scientific societies through the medium of their journals, of which the *Journal of Soil Science* (British), *Soil Science* and *Proceedings of the Soil Science Society of America* (American) have a wide circulation. *Soviet Soil Science*, a translation of *Pochvovedenie* is available for those interested in the activities of Russian soil scientists. The national survey organisations of most countries produce memoirs or bulletins and the World Soil Resources Office of the Food and Agriculture Organisation of the United Nations also publishes information on soils. Scattered articles on soils can be seen in geographical, ecological, agricultural and geological journals.

Chapter 1

Joffe, J. 1949. *Pedology*. Pedology Publications, New Brunswick, New Jersey, U.S.A.

Russell, Sir E. J. 1957. *The World of the Soil*. (New Naturalist Series.) Collins, London.

Chapter 2

Buckman, H. O. & Brady, N. C. 1960. *The Nature and Properties of Soils*. Macmillan, New York and London.

Comber, N. M. 1960. *An Introduction to the Scientific Study of the Soil*. Revised Townsend, W. N. Edward Arnold, London.

Jacks, G. V. 1956. *Soil*. Nelson & Sons Ltd, London.

Russell, Sir. E. J. 1954. *Soil Conditions and Plant Growth*. Longmans, London.

Chapter 3

Jenny, H. 1941. *Factors of Soil Formation*. McGraw-Hill, New York.

Chapter 4

Duchaufour, P. 1965. *Précis de Pédologie*. Masson & Cie, Paris.

Hallsworth, E. G. 1965. *Experimental Pedology*. Ed. Hallsworth and Crawford, D. V. Butterworths, London.

Kubiena, W. L. 1953. *The Soils of Europe*. Murby, London.

Northcote, K. 1965. *A Factual Key for the Recognition of Australian Soils*. C.S.I.R.O. Divisional Report 1960. 2nd Edition. Adelaide.

Robinson, G. W. 1949. *Soils, their origin, constitution and classification*. Murby, London.

Soil Survey Staff. 1960. *Soil Classification. A Comprehensive System. 7th Approximation*. U.S. Dept. Agric., Washington.

Chapter 5

Clarke, G. R. 1957. *The Study of the Soil in the Field*. Clarendon Press, Oxford.

Milne, G. 1936. *A Soil Reconnaissance through parts of Tanganyika Territory*. Memoirs Agric. Res. Sta., Amani. Reprinted in *J. Ecol.* 35, p. 192.

Perry, R. A. *et al.* 1962, *General Report on Lands of the Alice Springs Area, Northern Territory, 1956–57*. (Land Research Series 6.) C.S.I.R.O., Melbourne.

Soil Survey Staff. 1960. *Field Handbook*. Soil Survey of England and Wales. Harpenden, England.

Taylor, J. A. 1960. Methods of soil study. *Geography*, 45, p. 52.

See also *Memoirs of Soil Survey of England and Wales*. Harpenden. *Memoirs of Soil Survey of Scotland*. Macauley Institute, Aberdeen.

Chapter 6

Dimbleby, G. W. 1952. Pleistocene ice wedges in N.E. Yorkshire. *J. Soil Sci.* 3, pp. 1–19.

Fitzpatrick, E. A. 1956. An indurated horizon formed by permafrost. *J. Soil Sci.* 7, pp. 248–54.

Ganssen, R. & Hadrich, F. 1965. *Atlas zur Bodenkunde*. Bibliographisches Institut, Mannheim. (German text, but maps have key in English, French, Spanish and Russian.)

Smith, J. 1956. Some moving soils in Spitzbergen. *J. Soil Sci.* 7, pp. 10–21.

Tedrow, J. C. F. *et al.* 1958. Major genetic soils of the Arctic slope of Alaska. *J. Soil Sci.* 9, pp. 33–45.

Chapter 7

Avery, B. W. 1958. A sequence of beechwood soils on the Chiltern Hills, England. *J. Soil Sci.* 9, pp. 210–24.

Mackney, D. 1961. A podzol development sequence in oakwoods and heath in central England. *J. Soil Sci.* 12, pp. 23–40.

Muir, A. 1961. The podzol and podzolic soils. *Advances in Agronomy*, 13, pp. 1–56.

Muir, J. W. 1955. The effect of soil-forming factors over an area in the south of Scotland. *J. Soil Sci.* 6, pp. 84–93.

Stobbe, P. C. & Wright, J. R. 1959. Modern concepts of the genesis of podzols. *Proc. Soil Sci. Soc. Amer.* 23, pp. 161–4.

See also *Memoirs of Soil Survey of England and Wales*. Harpenden. *Memoirs of Soil Survey of Scotland*. Macauley Institute, Aberdeen.

Chapter 8

Aubert, G. 1962. Arid Zone Soils. A study of their formation, characteristics, utilization and conservation. *Arid Zone Res.* 18, pp. 115–37.

Aubert, G. & Boulaine, J. 1967. *La Pédologie*. Que Sais-je? Presses Universitaires de France, Paris.

Bunting, B. T. 1965. *The Geography of Soil*. Hutchinson, London.

Duchaufour, P. 1965. *Précis de Pédologie*. Masson & Cie, Paris.

Gerasimov, I. P. & Glazovskaya, M. A. 1965. *Fundamentals of Soil Science and Soil Geography*. Tr. A. Gourevitch. Israel Programme for Scientific Translations. Jerusalem.

Stephens, C. G. 1962. *A Manual of Australian Soils*. C.S.I.R.O., Melbourne.

Chapter 9

Cunningham, R. K. 1963. The effect of clearing a tropical forest soil. *J. Soil Sci.* 14, pp. 334–5.

Ellis, B. S. 1952. Genesis of a tropical red soil. *J. Soil Sci.* 3, pp. 52–52.

D'Hoore, J. L. 1964. *Soil map of Africa*, 1 : 5,000,000. C.T.C.A., Lagos.

Klinge, H. 1965. Podzol soils in the Amazon Basin. *J. Soil Sci.* 16, pp. 95–103.

McNeil, M. 1964. Lateritic Soils. *Scientific American* 207 (11), pp. 97–102.

Mulcahy, M. J. 1960. Laterites and Lateritic soils in South-western Australia. *J. Soil Sci.* 11, pp. 206–25.

Nye, P. H. 1954. Some soil forming processes in the humid tropics. 1. A field study of a catena in the West African Forest. *J. Soil Sci.* 5, pp. 7–21.

Watson, J. P. 1964. A soil catena on granite in Southern Rhodesia. 1. Field observations. *J. Soil Sci.* 15, pp. 238–50.

Chapter 10

Avery, B. W. 1958. A sequence of beechwood soils on the Chiltern Hills, England. *J. Soil Sci.* 9, pp. 210–24.

Crompton, E. 1952. Some morphological features associated with poor soil drainage. *J. Soil. Sci.* 3, pp. 277–89.

Chapter 11

Baren, F. A. van- 1960. Soils in relation to population in Tropical Regions. *Tijd. voor econ. en Soc. Geog.* 51, pp. 230–3.

Bartelli, L. J. *et al.* 1966. *Soil Surveys and Land Use Planning*. Soil Sci. Soc. Amer., Madison.

Bibby, J. S. and Mackney, D. 1969. *Land Use Capability Classification*. Technical Monograph No. 1. Soil Survey of Great Britain. Harpenden.

Bidwell, O. & Hole, F. D. 1965. Man as a factor of soil formation. *Soil Sci.* 99, pp. 65–72.

Klingebiel, A.A. & Montgomery, P. H. 1962. *Land Capability Classification*. U.S. Dept. Agric. Handbook No. 210, Washington.

INDEX

References to countries are listed alphabetically under continental blocks: Africa, America, Australasia, Eurasia